YOUR

BEHAVIOR

Understanding and Changing the Things You Do

YOUR BEHAVIOR

Understanding and Changing the Things You Do

Richard H. Pfau, Ph.D.

Paragon House

First Edition 2017

Published in the United States by
Paragon House
St. Paul, MN
www.ParagonHouse.com

Library of Congress Cataloging-in-Publication Data

Names: Pfau, Richard H., author.
Title: Your behavior : understanding and changing the things you do / Richard H. Pfau, Ph.D.
Description: First Edition. | Saint Paul, MN : Paragon House, 2017. | Includes bibliographical references and index.
Identifiers: LCCN 2016037557 | ISBN 9781557789273 (pbk. : alk. paper)
Subjects: LCSH: Human behavior. | Perception. | Behaviorism (Psychology)
Classification: LCC BF199 .P43 2017 | DDC 150--dc23 LC record available at https://lccn.loc.gov/2016037557

The paper used in this publication meets the minimum requirements of American National Standard for Information Sciences—Permanence of Paper for Printed Library Materials, ANSIZ39.48-1984.

Manufactured in the United States of America

10 9 8 7 6 5 4 3 2 1

To the memory of
William T. Powers,
the Hari Seldon of our time.

Contents

PART II
You and Your Behavior

Chapter 3

Chapter 4

List of Highlights

Preface

Many psychology articles, books, and theories are simply descriptions of behavior—summarizing and discussing what people do. Many are basically "just so stories," akin to Rudyard Kipling's "How the Camel Got His Hump," containing imaginary ideas and abstractions such as "egos," "cognitive maps," and "psychological forces" that the authors treat as if they really exist, rather than being simply figments of their and others' imaginations. When based on research, many psychological writings go way beyond the limited data gathered, a great deal of which is about college students, and overgeneralize to the point of producing even more "just-so stories" and fads—ranging from left-brain, right-brain recipes for success, to self-help books for improving your self-esteem and brain-power. In addition, most work and thinking in psychology is narrowly focused, dealing with limited, often specialized areas that aren't integrated with other thinking and work, and often don't recognize the potential usefulness and validity of other scholarly thinking and research done concerning human behavior. Most do not present holistic views of what people do or address important but less apparent influences upon their behavior.

This book is a reaction to the overall scattered and speculative state of psychology at the present time. It aims to synthesize work and thinking from psychology, sociology, anthropology, biology, cybernetics, and other areas into a coherent explanation of why you and other people do the things you do, in a way that can be understood by an average educated person as well as by professionals in the fields mentioned (for whom extensive endnotes are provided to aid their understanding and possible follow-up).

As you will see, nearly everything that you do is done automatically, without thinking, in a particular context, for a particular reason. Some things you do are also done consciously, but these are a small part of your behavior. Most things you do are done in an environment whose influence is so subtle and commonplace that you are hardly aware of it. In any case, whether you do things automatically or more consciously, whether you are mindful of your behavior and environment or not, you will soon learn why you do the things you do: usually in order to achieve perceptions that matter to you! You will also obtain step-by-step guidelines to help you better understand and change your behavior, if you really wish to do so.

—Richard H. Pfau
Mansfield Center, Connecticut

Acknowledgments

This book has been influenced by many people. The first to read the entire draft manuscript, comment upon it, and suggest improvements were Diane Smith and Bill Stinson, two members of Nepal 7, the Peace Corps group of which I was a part. Comments by Pujeeta Carter resulted in the rearrangement of several chapters. Suggestions by Mohit Manandhar included the loan of a book and insights into "motivational interviewing" that I found very helpful. Feedback from Raymond Pfau and Chandra Cranse was also helpful.

Members of the Control Systems Group also provided encouragement, useful feedback, and direction. Warren Mansell read an early version of the book and has been supportive ever since. Kent McClelland and Fred Nickols provided useful information about portions of the book that present some of their own thinking and writings about Perceptual Control Theory (PCT) concerning social interactions and implications of PCT for managers. Bob Hintz, John Kirkland, David Goldstein, Philip Yeranosian, Angus Jenkinson, and Fred Good provided valuable feedback that, among other things, led to extensive revisions of the chapter dealing with the analysis of behavior. Comments by Martin Taylor, Lloyd Klinedinst, and Boris Hartman were thought provoking and helpful. I was also encouraged by feedback from Alison Powers. Richard Marken's deep understanding of PCT as reflected in his postings and interactions on the Control System Group's Listserve and by his publications was also helpful and influential.

A discussion with Gina Panettieri, an agent attending CAPA University, a conference of the Connecticut Authors and Publishers Association, prompted me to add an Introduction to the book. Roberta J. Buland read and edited the entire manuscript, helping to improve almost every page before I provided it to the publisher. Fred Good provided a number of valuable editorial suggestions afterwards. Jake Fitzgerald's professional preparation of a number of illustrations is also appreciated and acknowledged. Also acknowledged are permissions by Martin Levinson of The Institute of General Semantics, Fred Nickols of Distance Consulting, and Debra Schafer of The University of Michigan Press, to use copyrighted materials. The free granting of permission by the Cambridge University Press and the American Psychological Association's permissions policy permitting limited use of figures and text extracts is also appreciated and acknowledged.

On the home front, my wife, Geeta, was supportive and patient during the entire 13 years that I conducted research for and prepared what you will read. Without her loving presence and backstopping, this endeavor would not have been possible.

Finally, I would like to recognize the significant influence that William T. Powers has had on my thoughts and actions. As you will see, his thinking and theory is highlighted throughout this book. Before I came across his writings, I had been dissatisfied with the state of psychology for many years, considering it to be still in the "Middle Ages." While conducting research to develop an improved model of human behavior, I discovered his work and realized that his model and theory satisfied my quest for significant improvement that was behind my efforts. My acknowledgment of his influence is profound and a reason why this book is dedicated to him.

Introduction

Have you ever done something and then asked yourself, *Why did I do that?* Or perhaps you see someone else doing something strange and wonder, *Why does she act that way?* Maybe you sometimes feel that you should do something differently and think, *I really need to change what I'm doing, but how do I do it?*

This book will help you to answer such questions by explaining how to do these three things:

- Understand why you behave as you do.
- Understand why other people behave as they do.
- Change your behavior if you really want to.

To do these three things well, you will learn how your body works to control what you experience, perceive, and do.

We'll begin by discussing what a good survivor you are and how you are organized to stay alive and survive. Then we'll discuss how you became the way you are and how regular and routine your behavior is—and isn't. We'll take a good look at what is around you and how your environment influences you, often in subtle ways that you are not aware of.

Based on that knowledge, you will know in a general way why you do the things you do. You will also know, generally, why other people do what they do too. Then we'll use what was discussed to help you better understand the reasons behind specific things that you do. For this, you'll be provided with a guide to help you systematically analyze behaviors of interest to you. By using the guide, you will be asked to consider and identify specific influences on each particular behavior that you are analyzing—preferably a recent or presently occurring behavior, rather than one that occurred long ago.

In addition, you will be introduced to a step-by-step way to change your behavior—an eight-step approach for changing the things you do, if you really want to. You'll also learn about resources to help you along.

As you will see, you can go into as much depth as you like to learn about and do these things. If a chapter really stimulates your interest and you want to know more about what is discussed, a list of further readings is provided. If you are a serious scholar, you may want to look at the endnotes, since they can help you learn more about what is being discussed. If you are a light reader and prefer to skim around more, the "Big Picture" at the beginning of each chapter will give you an idea what the chapter is about.

Therefore, you have some options when reading what follows. Feel free to skip around the book if you want to, and read the parts that are of most interest to you. However, if you want to understand your and others' behavior and learn how to change your behavior as well as possible, you are encouraged to read each of the chapters that follow. If you really want to understand, analyze, and change your behavior, read it all! By reading and considering the earlier chapters, you will have a stronger basis for analyzing and changing your behavior as described later in the book. You will also learn things that are at the cutting edge of psychology—ideas that some believe mark the beginning of a revolution in the way that we understand human behavior.

PART I

You and Your Body

The two chapters that follow focus on how your body affects what you do. As you will see, your body's structure and organization as well as how your body functions to preserve your life are key factors behind why you do the things you do.

CHAPTER 1

YOU ARE A SURVIVOR!

The Big Picture

Survival is an important part of what you do and why you do it. Ultimately, your survival is the main reason behind nearly all that you do, because if you make a serious mistake, you can die! *Why you do the things you do is mostly a result of your ability and urge to survive.* And the same holds for those around you and elsewhere in the world. Why you are a survivor is at the heart of this book.

Some Evidence and Key Ideas

You are a survivor. You are structured and organized to stay alive. So were your ancestors, from your parents on backwards to the furthest reaches of your ancestry. The same holds true for everyone else on the earth today. You and they have all survived long enough to exist at this moment, from the youngest child to the oldest person alive today.

You can't deny it. For example, whether you believe in "young earth creationism"—that the first man and woman were created about 6,000 years ago—or whether you believe in evolution—that the first tool-making, human-like people existed more than 2,000,000 years ago—you can't deny that your ancestors lived one after another from that time to the time that you were conceived in your mother's womb. If you believe in the more recent time of creation, that means about 300 generations of fathers and mothers lived on both sides of your family, back to the first woman and man created by God. If a belief in evolution is more to your liking, at least 100,000 generations of mothers and fathers lived on your mother's and on your father's sides—all of whom lived long enough to have children who are your direct ancestors.

That's a pretty good survival rate! Given the things that could have gone wrong, it is actually amazing. Just think, all of your ancestors lived long enough to have a child, who then lived long enough to have a child, and so on, until you were born. None of your direct ancestors died in infancy, starved to death, or died of thirst before conceiving a child. None were killed by disease, eaten by animals, ate anything poisonous enough to kill them, had a fatal accident, were killed by someone else, and died before having a child from whom you are descended.

As a survivor, you do things to survive. And you do those things quite well. However, your survival is not only because of you and what you have done. Others have also helped you and your ancestors survive. If you had no help after your birth, you would have died as an infant. But, your mother or someone else took care of you enough that you continued to live. You were fed, clothed, and kept warm and cool enough to live. You were cleaned and taken care of when sick. You learned many things from others that have helped you live a healthy life—from brushing your teeth, to washing your hands, and to being careful when you do things that may harm you. The same is true for your mother and your father, their mothers and fathers, and each of their direct ancestors.

Highlight 1.1 Survival is Your Default Mode!

"Not a single one of our ancestors died in infancy. They all reached adulthood, and every single one was capable of finding at least one heterosexual partner and of successfully copulating." (Dawkins, *River Out of Eden,* 1)

"...the possibility of death is always present, and survival...must be continuously won and re-won...."; "survival is conditional upon action...: failure to act successfully leads to death." (Binswanger, *Biological Basis,* 6, 91)

Many different views about the origins of man and woman exist. For example, Christian "young earth creationists" believe that humanity was directly created by God and that the earth and universe are less than 10,000 years old. "Intelligent design" believers indicate that divine intervention occurred at some point in the past, as evidenced by the complexity of life that exists.[1] Hindus, Polynesians, American Indians, Zulus, traditional Koreans, classical Greeks, the Norse, and other groups all have or have had their own beliefs about how and when mankind originated.[2] Datings by evolutionists go back over 2 million years to the origins of our presumed ancestors, *homo habilis.*

Regardless, however far back you go, whatever your belief about the origin of humanity, you have lived long enough to read these words. And your parents lived long enough to conceive you—at least 14 years or more. And your grandparents lived long enough to conceive your parents, and your great-grandparents lived long enough to conceive them, as did ancestors all the way back to the origin of humans. The same holds for the billions of other people on the earth and for all of their ancestors.

Thus, it is indisputable that you and other humans are survivors—and have been for many thousands of years.

Highlight 1.2 **Origin Beliefs**

Interesting examples of origin beliefs can be found online at *Wikipedia: The Free Encyclopedia,* under the headings "Creationism," "Theistic evolution," and "Creation myth."

Autopoiesis

What makes you such a good survivor? Your body's internal structure and organization as well as what you do are reasons you survive so well. Your body is arranged in a way that helps ensure your survival, insofar as survival is possible. What you do also affects your survival. However, what you do depends on your body and how it is structured and organized, as you will see later. So, let's consider your body more and how it relates to the idea of "autopoiesis"—a word that refers to the self-producing nature of your body.

Your body is extremely complex. Its many parts and systems work together to maintain your life and existence as a human being. For example, your heart pumps blood, oxygen, and needed nutrients to different parts of your body. Your skin protects you from the outside world and helps you keep cool by sweating. Your kidneys help remove waste products from your body and send them to your bladder where they are expelled. Your bones permit you to move by giving structure to your body. Your senses of touch, pain, temperature, vision, hearing, taste, smell, and balance help you avoid being harmed by your environment. If important parts of your internal mechanism break down, you die. But you haven't died yet, have you? Your complex body has worked for many years to maintain your life processes and, hopefully, will continue to do so for many more years.

However, your body depends on your environment for its existence and is affected by that environment in ways that can help you to live or that can kill you. Let's consider that aspect of your existence.

First, you depend on your environment for your survival. You need energy to live. Most of that energy is provided by food. You need water and oxygen and other chemicals such as salt to help convert the food that you eat into energy in order to build up your body's tissues and to help you function properly. That food and those chemicals are parts of your environment that you depend on to live. But your environment can also kill you. You can drown, fall, be killed by someone else or by an animal, eat something poisonous, become infected by a disease, be struck by a car, or be burned by fire.

However, every day you avoid dying by obtaining the nutrients and energy you need to live. You avoid being killed. You avoid becoming deadly sick or starving to death. How? Because your body is structured and organized to live. It senses your environment and reacts to it in ways that help you survive. Many of these ways are automatic. For example, if your hand is burned, you automatically pull it away from what burned it. If your body becomes too hot, you sweat to cool off. You breathe without thinking so that your body obtains the oxygen it needs to survive and expels harmful carbon dioxide. You don't eat rotten food: food that contains harmful germs that may make you sick and kill you. You don't eat it because it smells bad and you are repelled by it. Less obviously, your immune system protects you from viruses, bacteria, and parasites and helps you recover when you are sick. You react in these ways because your body is organized and structured to do such things naturally and automatically.

Other things that you do are learned. You learn not to step in front of a speeding car, how to obtain food, and how to keep warm and cool by wearing clothes or taking them off. You learn where to go to sleep at night, how to ask someone for help, and how to talk. When you learn such things, the structure of your body actually changes (as you'll see later in Chapter 4).

In other words, you are naturally structured and organized to sense, learn, and deal with your environment in ways that help you to survive. *You are a survivor*! How you are structured and organized to help you survive is one of the themes of this book.

Why you do what you do is based on your organization and structure as a living, surviving, *autopoietic* being—as Highlight 1.3 explains.[3] As you can see by looking at that Highlight, one characteristic of autopoiesis is that what you do is determined by your internal setup and not by your environment. You may react to things around you, but they do not force you to behave in a certain way. Your nervous system and the perceptions that result from its structure and organization are especially important parts of the process that results in your behavior. The same holds for other people too.

Highlight 1.3 Autopoiesis

You and other living organisms are autopoietic beings, according to Humberto Maturana and Francisco Varela, two scholars from Chile. An autopoietic being is a continually self-producing system. You continually maintain the living structure of your body and, by doing so, survive, as do all other living systems.

Some of your characteristics as an autopoietic being are these:

- What you do at any given moment depends on your internal organization and structure at that time. It depends on internal chemical processes, for example, between neurons, and between neurons and muscles. And this organization depends on (a) your genes and (b) your life history and experiences from conception to the present.
- The outside environment may trigger what you do. It may set into motion reactions on your part. But it does not cause you to do certain things or behave in specific ways. It does not specify what you do. What you do is determined by how you are structured and organized internally, not by outside events or objects.
- Focusing on neurons again, for example, "Another way of saying this is that the structure of the nervous system at a particular time determines both what can trigger it and what the outcome will be," according to John Mingers, *Self-Producing Systems,* 76.
- What you do is subservient to conserving the autopoietic (self-producing) structure and organization of your body.
- When a living system stops maintaining its autopoietic structure and organization, it dies and disintegrates. If you stop maintaining your autopoietic structure, you will die!

Sources: Maturana and Varela, *Autopoiesis and Cognition* and *The Tree of Knowledge*; Mingers, *Self-Producing Systems.*

Here's an example. Suppose that someone walking at night sees a group of men on the sidewalk directly ahead. To avoid possible trouble, he or she crosses to the other side of the street and continues walking. A second person seeing the men walks past them, indifferently, hardly noticing the group. A third person seeing them smiles broadly, walks up to them, shakes hands with each, and warmly says, "Long time no see!" While each of the three behaved differently, the group of men (the environment) was the same in each case. The environment may have triggered or influenced what each person did, but it did not cause them to react in a specific way. What each one did depended on how he or she perceived the situation. It was not the environment that caused the action—as we will discuss more in later chapters.

As you read this book, you may notice the ideas contained in Highlight 1.3 are reflected throughout.

Preview of the Next Chapter

The next chapter explains how you are structured and organized to be such a good survivor. As you will see, much of what you do occurs automatically to keep you alive and well. You will also begin to see how important your perceptions are to your behavior.

Further Reading

For an introduction to autopoiesis, two good books are:

- Humberto R. Maturana and Francisco J. Varela, *The Tree of Knowledge: The Biological Roots of Human Understanding,* rev. ed. (Boston: Shambhala, 1992).
- John Mingers, *Self-Producing Systems: Implications and Applications of Autopoiesis* (New York: Plenum Press, 1995).

Endnotes

1. Morris, *The Young Earth,* and Scott, *Evolutionism vs. Creationism.*

2. A comprehensive guide to origin beliefs is the two volume encyclopedia by Leeming, *Creation Myths of the World.* A more accessible reading in many libraries is Hamilton, *In the Beginning.*

3. The term autopoiesis is derived from Greek. *Auto-* means self and *poiesis* means creation, production, or making.

CHAPTER 2

YOU ARE ORGANIZED TO SURVIVE

The Big Picture

As mentioned in the previous chapter, you survive because your body is structured and organized to survive. Even though you are so fragile that you could die in a matter of days, minutes, or even seconds, you don't. Why? Because your body does things, mostly automatically, that keep you alive.

How fragile are you?

- If you fall from a great height or are hit by a truck going 65 miles an hour, you will die instantly.
- Without oxygen, you will die in a few minutes.
- Without water, you will die in a few days.
- Without food, you will die in about a month.
- If you get too hot or too cold, you may die.
- If you are badly burned or eat the wrong things, such as poison or spoiled food, you may die.
- If you are infected by germs, you can die.
- If you are attacked by a savage creature such as a large animal, criminal, or enemy, or if you are bitten by a poisonous snake, you may die.
- If injured and you lose too much blood, you will die.

Within your body, if the sugar in your blood falls too much, you will suffer confusion, delirium, convulsions, coma, and death. If calcium levels in your blood vary too much, convulsions and death will occur. Slight variations in the acidity of your blood have similar effects. If it is too high, coma and death occur;

if it is too low, convulsions and death. If your brain does not receive blood for more than eight minutes, you die.

So why haven't you died yet? Because *you are structured and organized to live, not to die*. Your body deals with internal and external conditions effectively. It does this so well that, after the many years since you were born, you are still alive and reading this book.

Of course, you are also dependent on your environment to survive, as you have been since you were an infant, and even before as a fetus in your mother's womb—but that is something we'll discuss later. For now, let's focus on your wonderful body.

Your Natural Survival Abilities

Your Need for Oxygen

You need oxygen to live. Sometimes you need more oxygen, such as when you are working, exercising, or walking. Then you breathe faster and deeper, automatically. Your heart also beats faster and harder. These actions increase the oxygen picked up by your blood that is sent to your tissues where it is needed the most. Your faster beating heart and breathing also help get rid of carbon dioxide, a waste product that results from your increased activity. Excess carbon dioxide can kill you if it is not expelled.

When you need more oxygen for your muscles to function, your body reduces the amount of blood sent to less critical organs not in immediate need, such as your stomach and intestines, and directs more blood to your muscles while maintaining the supply to your brain. How? Blood vessels going to your muscles dilate, become bigger, and carry more blood. With this extra blood and the oxygen it contains, your muscles perform more effectively—much better than if your body didn't respond to the need for more oxygen. Also, given its adequate supply of oxygen, your brain continues to function effectively.

Similarly, if your body's oxygen supply is cut off by your face being under water, unwillingly or for too long, you will struggle furiously to get to the surface, breathe, and survive. If you are not a swimmer, you will randomly thrash about in the water. If you are a good swimmer, you may control your movements in a less random manner.

If submerged, your body senses the situation in another way, too, and responds by what is called the "mammalian diving reflex." This reflex puts your body into an energy saving mode and increases the time it can stay under water without air. Your heart rate will slow up to 50 percent and the flow of blood to your hands and legs will be restricted so that the blood and oxygen flow to your vital organs, especially to your brain, will increase.[1] If water enters your mouth, nose, and throat, your vocal cords (larynx) will close so the water doesn't enter your lungs.

Differently, if you go up to a high altitude where there is less oxygen, over time your body will produce more red blood cells to carry more of the limited oxygen that is available. By doing so, you are able to function more normally than when you first arrived.

Obtaining sufficient oxygen happens automatically, because your body is structured to (1) sense a real or possible lack of oxygen, or a related increase in carbon dioxide, and (2) respond to that need or problem. And this is what happens with only one of the many critical things essential to your life—oxygen. Thanks to your wonderfully structured and organized body, many other automatic processes also happen at the same time and keep you alive.

Other Examples of Survival

Your body loses water when you breathe, sweat, urinate, have diarrhea, and in other ways. Loss of too much water results in an automatic response. You become thirsty, and eventually look for something to drink. If you need food for energy, your body indicates that too. You become hungry, uncomfortable, and eventually try to find something to eat.

Your body also maintains a relatively constant temperature—one that permits you to function well. This is important because, if your body becomes too warm, your cells will die. If you become too hot, you sweat, which helps cool you. If you become too warm, you may also do other things to cool yourself, such as removing a layer of clothing or turning on a fan or air conditioner. Also, becoming too cold can kill you. But when you become too cold you begin to shiver, and that muscle activity raises your body temperature. You may also put on warmer clothes or go to a warmer place.

Most of these things that you do to survive are done automatically. They are built into you as a result of your body's structure and function. However, others are learned: where to look for food to eat, taking off some clothes to help you cool, and putting on more clothes to get warm. Such learning often occurs automatically and unconsciously—something that we'll discuss more in Chapter 4.

Why do you do these things? Why do you become hungry and look for food, become thirsty and look for something to drink, sweat when you are hot, and learn how to do things that help you survive? You do these because your body senses a change from a state that is related to your survival, a change from an autopoietically stable state. When this occurs, your body is organized to respond in order to eliminate or reduce the difference being sensed. And, you respond either internally—such as by sweating, shivering, or having your heart beat faster, and/or by doing something external to yourself—such as by putting on warmer clothes or looking for food or water. These actions and behaviors help you to maintain your essential internal states. They help you to survive.

Automatic and More Conscious Processes

Internally, through its structure, organization, and related chemical actions, your body automatically maintains critical levels crucial to your survival such as blood sugar and calcium levels, blood acidity, body temperature, and the blood supply to your brain. The same goes for how you deal with what is outside your body—although, then, you may be more conscious of what is happening and try to consciously control what you do, rather than just doing things automatically.

Highlight 2.1 — Some Concepts

Structure refers to the physical parts of the body that can be seen, such as the shape and parts of the nervous system and bones.

Organization refers to the relationships between parts of the body, and how the parts work together and function, for example, the processes that occur as your senses, brain, and muscles work together and function in order to help you survive and achieve your goals.

Here are some familiar examples of your body's automatic processes:

- You are organized to have a fear or awareness of heights, which helps prevent you from falling down stairs and off other high places.
- When a foreign particle gets into your nose, you sneeze.
- If food or a drink goes the wrong way down your throat and lungs, you cough.
- If you eat bad or rotten food, you may vomit.

- If a small particle comes in contact with your eye, you will produce tears and blink to help wash it out.
- When infected by germs, your body's immune system automatically attacks them.
- When you have a wound, your blood will clot, bleeding will stop, and the wound will usually heal.
- If your skin is thin in an area used often that may break or wear away, you'll develop calluses or thickening of the skin.
- When you sense danger, your body has a natural fight or flight response that prepares you for quick action.
- If you haven't eaten for a long time, feelings of hunger stimulate you to obtain food and eat.
- If you haven't drunk anything for a long time, you experience feelings of thirst that stimulate you to drink fluids.

Your body is also structured and organized to automatically produce pleasant and unpleasant feelings: pain, discomfort, fear, pleasure, joy, and happiness. Interestingly, even these feelings help you to survive. You usually avoid objects, events, and behavior that produce bad feelings—things that are usually wrong for you, that produce pain, suffering, or discomfort. In contrast, you usually seek and do things that give you good feelings, such as eating a meal when hungry, drinking when thirsty, turning on a fan or air conditioner when hot, and seeking the company of a friend or romantic partner. These kinds of feelings provide quick and efficient guidelines for you to act and, when you have time, to help decide what action to take. These guides are especially helpful when time may be too short to think about what to do. For example, if you get a bad feeling about someone approaching you, you may rapidly try to avoid the person or prepare yourself to deal with him or her.

Thanks to your nervous system, you are structured and organized to sense your environment—to detect dangers and opportunities around you and to react to those dangers and opportunities. Your sense of smell, for example, helps you avoid dangerous foods and substances. You are repelled by the smell of rotten food, vomit, and feces—things that contain bacteria or toxins that could harm you. Your sense of vision helps you detect dangers and beneficial situations: cliffs and other dangerous high places, for example, and food in the refrigerator. Your sense of touch helps you avoid injuring yourself more, such as when you touch a sharp pin or burning object. It also helps you to make love and perform other activities that are pleasurable. Your senses of taste, hearing, and balance also help you deal with your environment.

You also have a number of automatic mechanisms within your body that help store materials that are needed for your existence. Extra water is stored in your muscles and skin. Your spleen stores extra blood that is released when needed. Fat cells store extra energy in the form of fat. Your liver can carry reserve protein. Calcium is stored in your bones.[2]

Highlight 2.2 — Darwin Award Winners

Examples of people not as good as you at surviving:

- The man killed by an express train when he put his ear on the track to hear if another train scheduled to stop at his station was coming.
- The fellow killed by a 900-pound Coke machine that fell on top of him as he tried to tip a free soda out of it.
- The man who played Russian roulette with an automatic pistol.
- The fellow who suffered partial decapitation when he looked into the mouth of a launching tube containing what he thought was a faulty aerial firework.
- The lawyer demonstrating the safety of windows in a Toronto skyscraper who crashed through the glass with his shoulder and fell 24 floors to his death.

Sources: Northcutt, *Darwin Awards: Evolution, Darwin Awards II,* and *Darwin Awards: Survival.*

Besides these automatic mechanisms, your body permits you to learn. As a result of how your nervous system works, you learn both automatically, without awareness, and consciously. You may not have thought much about it before, but you have actually learned many things that help you to survive. For example, you have learned to obtain food and water. You know where the refrigerator and faucet are, don't you? You have learned ways to keep warm and cool. You have learned to recognize and avoid danger: When is the last time you stepped in front of a speeding car or put your hand into a burning fire? Of course, as Highlight 2.2 indicates, not everyone has learned as well as you, or is as smart or fortunate as you. But you are a good learner, good enough to deal with opportunities for meeting your needs and avoiding the life-threatening situations that you face, or you wouldn't still be alive. You have learned because your nervous system enables you to learn and cope with your environment (something that we'll discuss more in a later chapter).

In short, your body is structured and organized to help you survive. The following sections will help you understand more about how this occurs.

Details of Your Body and Its Organization

The Autopoietic You

As you may realize by now, you are an autopoietic being—a continually self-creating system that (a) maintains the structure and organization of your body, and (b) changes and often improves that structure and organization by doing and learning new things. You continue living by maintaining a relatively stable internal environment—one having a fairly constant temperature, metabolism, and blood oxygen level. In order to maintain your autopoietic existence, you must, of course, also deal with your external environment, with what is outside yourself. If you do not deal with that external environment sufficiently well, if you do not deal with its dangers and the opportunities it contains to meet your basic needs, you will die and physically disintegrate. Fortunately, as we have discussed, your body is organized and structured so that it is adept at surviving. Let's look in more detail at some of the ways that it functions to keep you alive.

You Maintain a Stable Internal Environment: Homeostasis

Obviously, your internal environment is important to your life. If your heart stops beating, if you get too hot or cold, if the acidity of your blood is too high or low, you die. The same holds true for all beings.

The importance of our internal environment was highlighted by Claude Bernard, a French physiologist, around 1865. Barnard concluded that stability of our internal environment (the *milieu intérieur* as he called it) is necessary for the maintenance of life.[3] However, it was not until the 1920s and 1930s that Walter Cannon, another physiologist, clarified such stability and coined the word *homeostasis* to describe the mechanism by which it occurs. Cannon stated that:

> The coördinated physiological processes which maintain most of the steady states in the organism are so complex and so peculiar to living beings...that I have suggested a special designation for these states,

homeostasis. The word does not imply something set and immobile, a stagnation. It means a condition—a condition which may vary, but which is relatively constant.[4]

Highlight 2.3 A Model of Homeostasis

The following diagram indicates how the homeostatic systems of your body work.

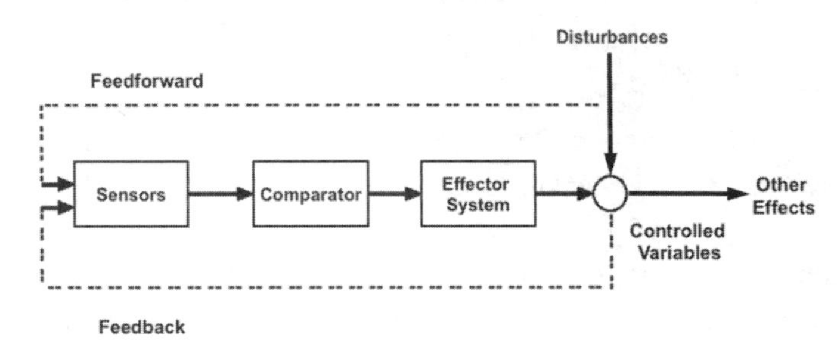

Controlled variables refer to things controlled, such as the oxygen supply to your muscles or the water balance of your body. The *feedback* loop results in sensors registering the status of the controlled variable. The *comparator* determines if there is a difference or error between what is sensed and what should be sensed. If there is a difference, the *effector system* (such as your blood-oxygen transfer system or water maintenance system) is activated so that the controlled variable is kept at or near its target level, with the feedback loop indicating the status of the controlled level. Activation continues until the controlled variable is at the target level. For example, if you lack oxygen, you will breathe faster and deeper and your heart will beat faster until your sensors indicate that the target level has been achieved; or in the case of water, you become thirsty, look for, and drink water. *Disturbances* also act on your homeostatic systems and levels of controlled variables. For example, increased muscle activity affects the levels of oxygen needed by your muscles, raises your body temperature, and may result in water loss through sweating. In some cases, feedforward processes activate sensors, comparators, and the effector system so that your body makes adjustments either (a) before the disturbance has an effect or much of an effect on the homeostatic system and controlled variables or (b) even before the disturbance occurs, such as when the disturbance and its effects are anticipated. This is what happens, for example, when you anticipate danger and your fight or flight response is activated, and when you do things such as put on a coat before going outside during cold weather.

Source: Adapted from Houk, "Homeostasis and Control Principles."

Cannon recognized the need for steady states in the body and described a number of mechanisms by which stability is achieved, including those by which the temperature of the body is stabilized and sufficient oxygen is supplied to tissues, as briefly described before. Other scholars, focusing on the fact that many steady states being maintained may vary over time or in different conditions, created words to describe such changes—words such as *homeodynamics* and *rheostasis*. These refer to systematic changes in the reference levels or conditions maintained over time, such as the temperature of the body, which is lower at night and higher during the day, changing hormone levels of women during their monthly menstrual cycles, and changes that occur during puberty. The mechanisms that maintain such stability have been clarified by scholars using models they call "control systems." One model is shown in Highlight 2.3 (opposite page).

Many homeostatic systems are completely internal, physiological processes. However, others, such as thirst and hunger, also have visible behavioral components that require you to act on your environment to maintain your internal homeostatic states.[5] For example, as a baby, when hungry, you cried to obtain food. As an adult, when hungry, you may have to cook a meal or order food at a restaurant to satisfy your hunger pangs. So, let us now look at what you do when you take action to deal with the world outside of your autopoietic body.

Highlight 2.4 A Few Definitions

Homeostasis: the regulation by an organism of all aspects of its internal environment. This involves monitoring changes in the external and internal environments by means of receptors and adjusting bodily processes accordingly. (VandenBos, *APA Dictionary, 2nd ed.*)

Homeodynamics: refers to the fact that many reference levels, or "set points" as he calls them, change during an organism's lifetime. (Rose, *Lifelines, 157*)

Rheostasis: a condition in which, at any one instant, homeostatic defenses are present but over a span of time, there is a change in the level that is defended. (Mrosovsky, *Rheostasis,* 31)

You Cope with Your External Environment

Besides maintaining a stable internal environment essential to life, you need to deal with what is outside your skin, your external environment. To survive, you have to obtain essential things and assistance from "out there"—such as food, water, air, certain chemicals, and help from other people. You also need to deal with your environment to avoid being harmed. Besides being a source for what is needed to live, your environment also contains threats to your existence: potential dangers such as extreme temperatures and weather, electricity, and careless drivers. So, when dealing with what's around you, you must be able to do at least two things: (1) make use of opportunities that exist in your external environment that help provide the necessities of life such as food and water, and (2) avoid or cope with dangers that exist in your environment, dangers that may result in your injury and death. In other words, you must be able to behave well enough at the proper times and places in order to live, survive, prosper, and possibly reproduce.

Highlight 2.5 Some Protective Functions of Your Body

The process of metabolism: maintaining an adequate oxygen supply, suitable body temperature, balance of internal chemistries, and regulation of blood pressure and flow.

Basic Reflexes: sneezing, coughing, gagging, vomiting, producing tears, blinking, yawning, and vocal crying when an infant.

The Immune System: coping with viruses, bacteria, parasites, and toxic chemical molecules.

Slower Processes: callous on skin, wound healing, blood changes at high altitudes.

Feelings and Emotions: pain, fear, anger, rage, disgust.

Drives and Motivations: hunger, thirst, temperature control, pain avoidance.

Sources: Cannon, *Wisdom of the Body;* Damasio, *Looking for Spinoza;* and Provine, *Curious Behavior.*

As indicated before, your body has a number of organs, systems, and related processes that enable you to react appropriately to your environment—to do the right things at the right times in order to live and maintain your autopoietic structure and organization. Some actions are carried out automatically, often unconsciously, as indicated by Highlight 2.5. Others are carried out more consciously and deliberately—such as applying for a job or filling out your tax return. Let's look at some of the structures and processes behind these behaviors.

When you do the things you do, you usually use your muscles. Your muscles enable you to move your body and affect things in your environment—to walk, talk, and to touch and hold objects. Your muscles even enable you to look at things by moving your head and eyes, focusing the lenses of your eyes, and holding your head erect if you are standing or sitting. Your skeleton is also important: without bones, you would be a quivering mass of flesh. Both your muscles and skeleton are your main mechanical structures for doing things, for moving your body or parts of your body. But what movements should you make? How do you know what to do? Let's consider this for a moment. Let's start to think about how and why you move appropriately.

You are an autopoietic being. Your body is humming along well—at least well enough for you to be alive. Your internal systems are working fine and keeping you stable. So let's consider your environment. How do you know what is outside your body? Where is food that you can eat? Where is your bed? What is the weather like? You know these things based on your senses—such as your sight, taste, smell, and hearing—and from what you perceive through them. In fact, *everything you know about the present and the past is based on your senses and perceptions based on those senses.*

Your senses permit you to determine what is "out there." Sight permits you to see what is around your body. Hearing helps you determine what is happening around you and is essential to verbal talking and conversation, these being behaviors helpful for survival and a social life that most people crave. Your ability to taste helps you detect useful substances such as food and to detect harmful substances such as toxins, which are often bitter. Smell helps you avoid contact with harmful substances such as rotten food and attracts you to areas of delight such as a delicious meal or a potential spouse. Touch and sensations associated with touch, such as pressure and pain, also help you avoid or reduce danger and deal with objects in appropriate ways.

Your senses, of course, are a part of your nervous system, which coordinates your body's actions when dealing with the environment. Everything you do involves your nervous system, including your brain and the signals it sends to different parts of your body, such as your muscles. Your nervous system is important and is often referred to in this book. For now, though, think of your nervous system and brain as structures that enable you to sense and perceive your environment, act quickly, learn, remember, and imitate what others are doing. All these actions and abilities help you survive.

Highlight 2.6 Your Fight or Flight Response

The fight or flight response is also called the "stress response" or the "general adaptation syndrome." It is a response to physical or psychological conditions that either disrupts your homeostatic balance or may disrupt that balance, resulting in a perception of a threatening or otherwise stressful situation. The threat or stress results from a difference between your perception of a controlled variable and the way it should be. The response refers to your body's initial efforts to maintain or reestablish the target level of the controlled variable.

The response is fairly nonspecific. That is, many different situations or disturbances result in a similar response. As Randy Nelson, an endocrinologist, explains: "Two endocrine systems, one involving epinephrine (adrenaline) from the adrenal medulla and the other involving glucocorticoids from the adrenal cortex [of the brain], constitute the major components of the stress response.... Within seconds of perceiving a stressor, the sympathetic nervous system begins to secrete norepinephrine, and both adrenal medullae begin to secrete epinephrine. A few minutes later, the adrenal cortices begin to secrete glucocorticoids." [672] The hormones released cause changes in respiration, heart rate, blood pressure, blood glucose levels (for energy), and alertness, as well as enhanced learning and memory.

Adaptive effects, according to Nelson, include:

1. Increased immediate availability of energy
2. Increased oxygen intake
3. Decreased blood flow to areas not necessary for movement
4. Inhibition of energetically expensive processes that are not related to immediate survival, such as digestion, growth, immune function, and reproduction
5. Decreased pain perception
6. Enhancement of sensory function and memory.

Such a response usually helps improve your chances of survival by preparing you to deal with the situation being faced, thereby helping to maintain or reestablish your homeostatic balance or achieve an important preferred state. The response also helps you to learn and remember the situation and what you did. However, a prolonged stressful response as well as short but frequent stress responses can result in health problems.

Sources: Nelson, *Behavioral Endocrinology,* 3rd ed., 672, 682; Pavloski, "Physiological Stress."

Another set of structures that help you survive is your endocrine system. This system consists of glands that release chemicals called hormones into your blood and other bodily fluids to affect what happens locally, such as between nerve connections, and more distantly and broadly, such as by activating your fight or flight response. Whereas your nervous system excels in rapid action, your endocrine system tends to act more slowly, although some hormones such as adrenaline may act fairly quickly, such as occurs during the fight or flight response described in Highlight 2.6. Hormones help you to deal with everything from children, potential mates, and external dangers, to internal matters such as memory, tissue development, and, if you are female, to uterine contractions during birth and your ability to provide milk to a baby.

Perception and What You Do

As indicated before, when you deal with your environment, you only know what you sense and perceive. Your body reacts to your environment based on what you see, hear, touch, taste, smell, feel, and otherwise sense—including those hunger pangs we talked about before, thirst, and good and bad feelings. How you behave is affected by your senses and how you process the signals from your senses.

When you perceive something to be not quite right, something different from what you would like it to be, such as when you are hungry or see a TV program you don't like, you may do something to change that perception. For example, you may eat some food or change the TV channel to one with a program you like more. By doing so, you control and change your perceptions, usually to something better or more satisfying. Similarly, if you are driving and see your car drifting off the road, you will probably steer it back to the middle of your lane. At work or at home when cooking or when raising children, you may perceive that something needs to be done and then take action until things are as they should be—perhaps by completing a job task at work, by adding more salt when cooking, or by feeding a hungry child.

Alternatively, when you perceive that things are OK, you may continue doing what you are doing in order to maintain those perceptions. For example, if you feel the room temperature is comfortable, you probably won't change the thermostat setting of the heating or air conditioning system. If you like what you are eating, you may eat more of it. If you are comfortably asleep in bed, you may stay there unless you perceive a need to go to the bathroom and empty your bladder—in which case you will do what is needed until the sensations of your bladder are more satisfactory.

In short, *when you do something, you do what you do in order to change or maintain your perceptions*. This is a very important point, recognized by William T. Powers by 1960.[6] As Powers has stated so well, "*Behavior is the process by which we act on the world to control perceptions that matter to us.*"[7] In turn, as you will see in the next chapter, this is done to achieve goals or other reference states that you have—states that, for the most part, help you to survive and maintain your autopoietic being.

So, in addition to autopoiesis, the relationship between perception and behavior is another key notion behind what you do—one that we'll discuss in more detail in the next chapter under the topic "Perceptual Control Theory." But for now, keep in mind that your perceptions affect what you do! And, as indicated by the discussion of autopoiesis in Chapter 1, your perceptions and behavior are a result of how your nervous system, muscles, and overall body are structured and organized.

Highlight 2.7 — Perception and Behavior

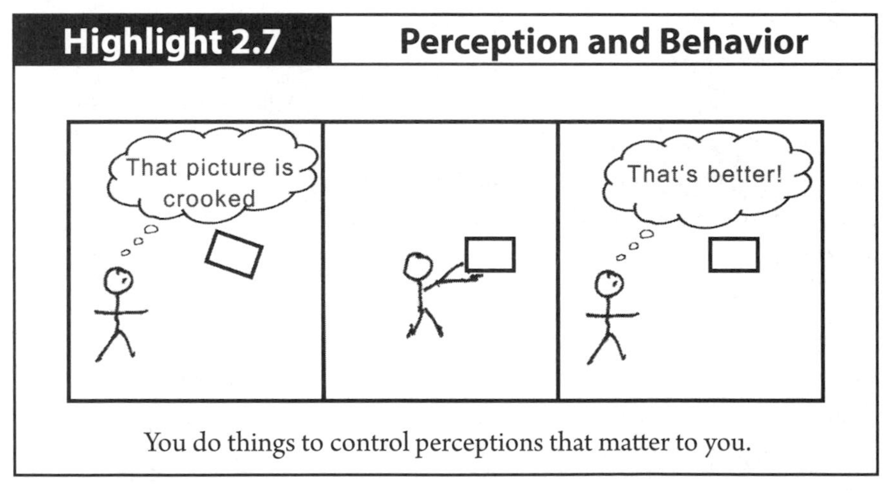

You do things to control perceptions that matter to you.

Preview of the Next Chapter

The next chapter, the beginning of Part II, goes into more detail about things that affect your behavior, especially those internal to your body. Part III focuses on factors external to yourself.

Further Reading

For information about homeostasis:

- Walter B. Cannon, *The Wisdom of the Body* (New York: W. W. Norton, 1932). The classic description of homeostasis in humans by the man who coined the term.
- Nicholas Mrosovsky, *Rheostasis: The Physiology of Change* (Oxford: Oxford University Press, 1990). A reexamination of homeostasis including changes in regulated levels.
- James C. Houk, "Homeostasis and Control Principles," in *Medical Physiology*, vol. 1, 14th ed., ed. Vernon B. Mountcastle (St. Louis: C. V. Mosby, 1980), 246-267. A good introduction to homeostasis and human control systems.

For information about the nervous system:

- Eric R. Kandel et al., eds., *Principles of Neural Science*, 5th ed. (New York: McGraw-Hill, 2013). A comprehensive, technical text.

For information about the endocrine system:

- Randy J. Nelson, *An Introduction to Behavioral Endocrinology*, 4th ed. (Sunderland, MA: Sinauer Associates, 2011).

Endnotes

1. This reflex is also called the Mammalian diving response" as Panneton does in his technical article "The Mamalian Diving Response" that describes physiological aspects of the response.

2. Cannon, *Wisdom of the Body*.

3. I. Cohen, "Foreword".

4. Cannon, *Wisdom of the Body*, 24.

5. Nelson, *Introduction to Behavioral Endocrinology*, 4th ed., 454-455.

6. Powers, Clark, and McFarland, "A General Feedback Theory: Part I" and "A General Feedback Theory Part II."

7. Powers, *Making Sense of Behavior*, p. 17. Note that all references to "Powers" when cited in the text and endnotes refer to William T. Powers. His wife, Mary Powers, when cited will be referred to as M. Powers.

PART II

You and Your Behavior

The three chapters that follow focus on the things you do—that is, on your observable behavior. We will discuss how internal processes and structures of your body affect what you do, go over how you became the way you are, and consider how routine and regular your behavior seems to be.

CHAPTER 3

WHAT AFFECTS WHAT YOU DO? A FOCUS ON YOUR INTERNAL PROCESSES AND STRUCTURES

The Big Picture

As you should realize by now, how your body is structured and organized affects your behavior as does what is outside your body, your external environment. Both your insides and your outsides are important.

The internal structures of your body that affect what you do are (1) your nervous system, (2) your endocrine system and its hormones, and (3) parts such as your muscles, skeleton, heart, lungs, and other organs. For example, your muscles and skeleton enable you to physically move in certain ways. However, your nervous system is especially important when thinking about what you do and why you do it.

The nervous system determines what is sensed and perceived, felt, learned, and remembered, including all that you know, believe, and can do. This is important because, first, you are not able to deal with things that you cannot detect by means of your senses. Second, your feelings are important because you usually do what feels good or seems agreeable and avoid doing things that are unpleasant. Finally, without learning and memories of past experiences and actions to guide what you do, your behavior would be like that of an insect; it would depend on what was built into you, on your reflexes and random actions. You would have no learned knowledge, skills, or abilities. Moreover, your nervous system coordinates your actions so that you can effectively deal with what is around you—from doing things such as walking and talking to working and making love. If you doubt this, consider people whose nervous systems are not functioning properly, such as those who are blind, paralyzed, or have Alzheimer's disease. Although they can do many things, they are limited in other ways.

Your hormones also coordinate and affect what you do, but their action is usually slower, partly because they are carried around your body by your bloodstream. Your hormones do not *determine* what you do. Instead, they work subtly by *increasing the chances* that you may do something, such as fight, run away, or be attracted to someone.

In addition, what you do is affected and limited by your body's other systems. For example, try performing well when you have a fever of 104°F (40°C) or running faster than a speeding bullet even when you are healthy.

What is outside your body, your environment, also affects what you do, especially objects, people, settings, and events that are important to you. Steps, doors, advertisements, cars, roads, traffic lights, trees, animals, food, and water all affect your behavior. Similarly, family and friends, teachers and bosses, police and strangers do too. So do settings such as schools, churches, workplaces, and stores as well as events like holidays, religious services, and examinations—where certain types of behavior are expected and considered appropriate.

But, as mentioned before, how such external things affect you depends on how you sense and perceive them—which, in turn, is based on the unique structure and organization of your body, including your nervous system. That structure and organization is what we will focus on in this chapter.

Your Nervous System

Your nervous system consists of cells called "neurons." That system is the structure that coordinates what you do in order to deal with the world around you. It is also the seat of your learning and memories. Major parts of your nervous system are your brain, sensory neurons, motor neurons that activate your muscles, and neurons that connect these parts. Every sensation, memory, and thought you have and everything you do is a result of your nervous system and how it operates. So let's look at it more closely.

Your Brain

Your brain consists of billions of neurons connected to one another in ways that enable coordination of your body. Coordination is achieved, in part, by receiving neural signals from your senses that make you aware of your internal and external environments, and by sending signals to your muscles and organs. The signals then cause you to move or otherwise deal with those environments—often automatically.

Your brain is also the organ that permits you to learn, think, remember, feel, imagine, and have goals and plans—all of which affect what you do. For example, your memories help you to act in appropriate ways based on your past experience. Decisions that you consciously make are a result of your thought processes and related memories. Your feelings, good and bad, also guide what you do in important ways that you are often unaware of. Your imagination helps you to visualize things that you may wish to do or avoid. Your goals and plans also guide your actions as do your values, obligations, wishes, and desires. All of these are a result of your brain's operation and processes.

Composed of many "control systems," the general organization and functions of which you'll learn about soon, your brain controls most of what you do. For example, if you are like most people, some of these control systems affect something that you do about one-third of your life every day or night. Can you guess what that behavior is? See Highlight 3.1 to find out! For now, though, an important point is that *your brain plays a key role in what you do.*

Highlight 3.1 The Sleep-Wake Cycle

Although sleep is one of your most common behaviors, its functions and mechanisms are poorly understood. You and other animals have a daily sleep-wake cycle called the circadian rhythm that is linked to the 24-hour cycle of night and day. About one-third of your time is spent in the sleep phase of the cycle. If you sleep much less than eight hours a day, chances are that you are sleep-deprived. This condition often occurs in societies because of artificial lighting; options for pleasurable entertainment including TV, computer games, parties, and movies; and work demands.

At certain times of the day, you may become sleepy. This is a natural occurrence, built into the systems of your body. Although the mechanisms are not fully known, a critical center for this cycle is located in a small part of your brain containing about 20,000 neurons called the suprachiasmatic nucleus. This area acts like a master clock for the cycles of arousal and sleepiness that you experience every day.

Sleep is important. Lack of sleep can cause death even more quickly than a lack of food!

Sources: McCormick and Westbrook, "Sleep and Dreaming," 1140-1158; Weaver and Reppert, "Circadian Timekeeping," 931-957; and Martin, *Counting Sheep,* 3-4, 76, 114.

Your Senses and Perceptions

Your senses are an important part of your nervous system. Let's consider them and your perceptions a bit more. How do you know what is happening inside your body—if anything is hurting you, or if you are hungry, thirsty, tired, ill, or in love? How do you know what is outside of your body—the dangers and opportunities that may harm or help you, such as speeding cars and sources of food and water? The answer is by means of your senses and related perceptions. All that you know about the world and your body is a result of your senses and resulting perceptions.

You cannot be aware of things or react to them unless you can detect them. Your sensory neurons enable you to do this—to react to light, sound, food, chemicals, objects, and other matter that touches, presses against, or injures your body. Sensory neurons also enable you to sense pain, temperature, and the position of your body. See Highlight 3.2.

Highlight 3.2 More Definitions

Sense: Any of the media through which one perceives information about the external environment or about the state of one's body in relation to this. They include the five primary senses: vision, hearing, taste, touch, and smell, and also the senses of pressure, pain, temperature, equilibrium, and kinesthesis, the sense of body position and movement. Each sense has its own receptors, responds to characteristic stimuli, and has its own pathways to a specific part of the brain.

Perception: 1. The process or result of becoming aware of objects, relationships, and events by means of the senses, which include such activities as recognizing, observing, and discriminating. These activities enable organisms to organize and interpret the stimuli received into meaningful knowledge. 2. A perceptual signal inside a system that is a continuous analogue of a state of affairs outside the system.

Sources: VandenBos, APA Dictionary of Psychology, 683, 836 and Powers, Behavior: The Control of Perception, 286.

External and internal events and conditions activate your sensory neurons. Impulses then move from one end to the other of those neurons and activate other neurons, including those in your brain, with the eventual result that you have a sensation such as of light, sound, and touch. Actually, however, all that is communicated are neural impulses of varying frequency and timing. Sometimes those perceptual impulses result in immediate, local, automatic action. However, in most cases your brain receives the impulses and then, based on its organization and processing, sends other impulses that affect what you do. For example:

- If your body has too much carbon dioxide (CO_2), a waste product in your bloodstream, what happens? A homeostatic process takes place whereby the excess CO_2 is sensed. Your brain and nervous system then stimulate your respiration and heart so that you breathe faster and deeper, and your heart beats faster, so that the CO_2 is expelled and reduced to more comfortable, preferred levels within your body.
- If your skin is poked by a pin or other sharp object, you perceive pain and act to reduce the pain, perhaps by withdrawing your body from the object or the object from your body.
- If you are playing baseball and a ball is coming towards you, if you are alert you sense the ball, perceive its movement, and try to catch or hit it.
- If your body needs water, you have a physiological sensation of thirst and then do something to reduce the thirst: If you are a baby, you cry; if you are older, you have learned how to find water, or ask someone for a drink.
- If you are driving a car, you sense where the car is and make adjustments so that it usually stays in the middle of the lane and moves at an appropriate speed towards your destination.
- If someone asks you to go to a movie with them, you sense and perceive the request and then react to it.

In each case, your senses and related perceptions permit you to become aware of and deal with your internal and external environments. In fact, you would not survive very long without your senses and perceptions. However, you do not act on every perception you have. For example, you may perceive that the leaves of trees are green or water is clear, but that perception does not necessarily affect your behavior. You only act when your perceptions differ in an important way from a reference that you have—as is discussed next.

Highlight 3.3 Perceptions Known to Affect Behavior

Perceptions of characteristics, opportunities, constraints, benefits, threats, feelings, and internal states shown by research to influence behavior include:

- **The attributes of another person:** Is the person friendly, honest, hostile, or a stranger, a thief, a banker, a doctor, a relative, a neighbor?
- **Appraisal of your context and situational cues:** Are you in a church, in a store, at a party, stopped by a policeman, in a burning building, walking in the woods, holding a hamburger or rat poison, in the path of an approaching car?
- **The expectations of others:** What do people expect you to do, especially those nearby who are a part of a situation you are in?
- **Inferences about social situations:** Might you be harmed or helped to do or achieve something you value? Is the person trying to sell you something?
- **Tastes and smells:** Such as bitter tastes, rotting smells, and delicious food.
- **Perceived efficacy:** Your confidence in your ability to do something.
- **Affective reactions and feelings:** Do you like or dislike what you are experiencing? Do you feel positively or negatively about the person, object, or situation, or about something you may do? Are you feeling angry, sad, depressed, disgusted, happy, confident?
- **Physiological states:** Hunger, thirst, sleepiness, being too warm or too cold, illness, pain, pregnancy.

Your References: Goals, Plans, and Other Referents

Just as your homeostatic states function around certain set points, your body acts to achieve or maintain other references. These references are often described by words such as "intentions," "goals," "objectives," and "plans."[1] People in technical or specialized areas such as science, business, and the military often use other words as Highlight 3.4 indicates, but all of these refer to a "reference"—a term that will often be used from now on.

Some of these references are natural states built into your body, such as your normal temperature, blood oxygen level, and calcium level in your blood. Some references are learned but fairly automatic, such as brushing your teeth in the morning or wearing clothes when you go outside in public. Some are quite conscious, such as perhaps intending to get a job or keeping a promise that you have made.

Highlight 3.4 **Words Indicating References**

Common Terms: goal, purpose, objective, intention, intent, desire, want, need, aim, value, preference, ambition, reason, concern, priority, yearning, aversion.

Social Terms: duty, moral or legal obligation, assurance, promise, pledge, commitment, solemn word, vow, word of honor, principle, norms, values, ideals, ideology, expectations of others, role, responsibility, chore, lawful/legal behavior, traditions, taboos, rules of behavior, commandments, mores, worldview.

Terms Referring to Sequences: standard procedure, modus operandi, script, schema, program, musical/dance score, choreography, critical path, recipe, program.

Scientific Terms: reference signal, reference value, reference level, reference condition, reference state, referent, set point, set zone, command signal, settling point, equilibrium point, stable state, preferred state, desired state, goal state, optimal outcome, optimal level, internal standard, function, display rules, phobia.

Business and Military Terms: target, mission, task, plan, blueprint, standard, assignment, strategy, benchmark, criteria, checklist, schedule.

Whatever you call these references, they do affect your behavior. For example, your goals and plans affect what you do: Haven't you ever planned a vacation or what to do during a summer break? Your intentions also affect your behavior. Did you ever intend to see a movie or look at a TV program and then do it? Your values affect what you do too. Perhaps you believe in being an honest person or, alternately, do not value honesty at all, or only act honestly in situations when your goal is not to be caught. Even though such references often affect what you do, sometimes they do not. You may decide not to act on your goals, plans, intentions, and values. Why? Some reasons include a lack of opportunity to do what you want to do, distractions, and your feelings, which may be stronger toward achieving another reference and doing something else.

The importance of references and the processes that lead to action based on those references has been recognized by cyberneticists for some time. Focusing on human behavior, "perceptual control scholars" first led by the late William T. Powers, point out that our references are routinely compared to our perceptions. When there is an important difference between what we perceive and our references, we take action when able, in order to reduce the difference. In other words, *you control your perceptions of the external world and your internal states, as much as possible, until they agree with your references.* You do this especially when your references are directly related to your survival, like having enough oxygen, food, water, sleep, and maintaining a relatively constant body temperature.[2]

According to Powers, we do things to reduce differences between our perceptions and references until the differences, or "errors" as he calls them, are eliminated or small enough to be ignored. We do this unless doing so (1) interferes with or conflicts with a more valued reference, (2) the error really does not make much practical difference, or (3) we think we aren't able to do anything to reduce the error or achieve the reference.

Here are some examples:

- You are hungry (a perception indicating a difference from your reference of not being hungry). As a result, you look for food and eat until you are satisfied (a perception that matches your reference).
- You want to look good for a job interview (a reference). The clothes that you are wearing are not suitable (a perception resulting in a difference from your reference), so you put on other clothes (your behavior) until you perceive that you are dressed properly (a perception that results in little or no difference from your reference for looking good).
- You see a red traffic light (a perception) while driving and moving toward it (a perception that differs from the legal reference that requires a car to stop in front of a red light). You step on the brake (behavior) until you slow down and stop (resulting in a perception that meets the reference specified by the law).
- You want to see a popular movie (a reference that differs from your perceived present state of not having seen it) so you go and see the movie (these being behaviors that result in perceptions that meet your reference).

We will discuss Powers' ideas more later in this chapter, but for now you hopefully realize that your references affect what you do. Although some of these references are genetically built into you, such as your blood oxygen levels and a desire for non-painful conditions, others are learned, such as your values, social customs, and many of your preferences. All these references result from how your nervous system is organized—another topic we will look at more closely later. But whether intrinsic or learned, your references are an important part of the process affecting what you do.

Your Neural Action System

To do things, you also need a "neural action system" to move your body. This system consists of your brain and the nerves that activate your muscles, as well as the sensory neurons that let you know what you are doing. The neural action system not only activates your muscles, but it also activates them in a suitable way for the task at hand based on your past experience and resulting perceptions. Although some of this function is genetically determined and built into you, much is learned. How such learning occurs and is expressed in your thinking and behavior is discussed in Chapter 4.

For now, remember that your brain and its linkages to your muscles affect what you do.

Conscious Thought

Based on your brain's activity, you also consciously affect some of your behavior, although you do this relatively rarely in comparison to how your body automatically operates. You consciously do things, especially (1) when you are facing a new or unexpected situation, (2) when you are trying to learn a new behavior, (3) when you are trying to improve a skill that you already have, and (4) when you realize that an automatic, habitual, or routine response that you are making (or may make) is not appropriate.[3] Otherwise, most of what you do is done automatically, with little or no conscious thought.

Highlight 3.5 — Consciousness

"...the devices of consciousness handle the problem of how an individual organism may cope with environmental challenges not predicted in its basic design such that the conditions fundamental for survival can still be met." (Damasio, *The Feeling of What Happens,* 303)

Models of What You Do

Based on what we have discussed so far, the following models will help you to better understand why you do the things you do.

Perceptual Control Theory

As indicated before, Perceptual Control Theory (PCT) is the insightful product of William T. Powers. This theory focuses on the idea that we do things until our perceptions are more in keeping with our references. In other words, as Powers puts it, "Behavior is the control of perception." The theory represents what some professionals consider to be a revolution in psychological thinking analogous to the revolution in biological thinking started by Darwin's theory of evolution.[4]

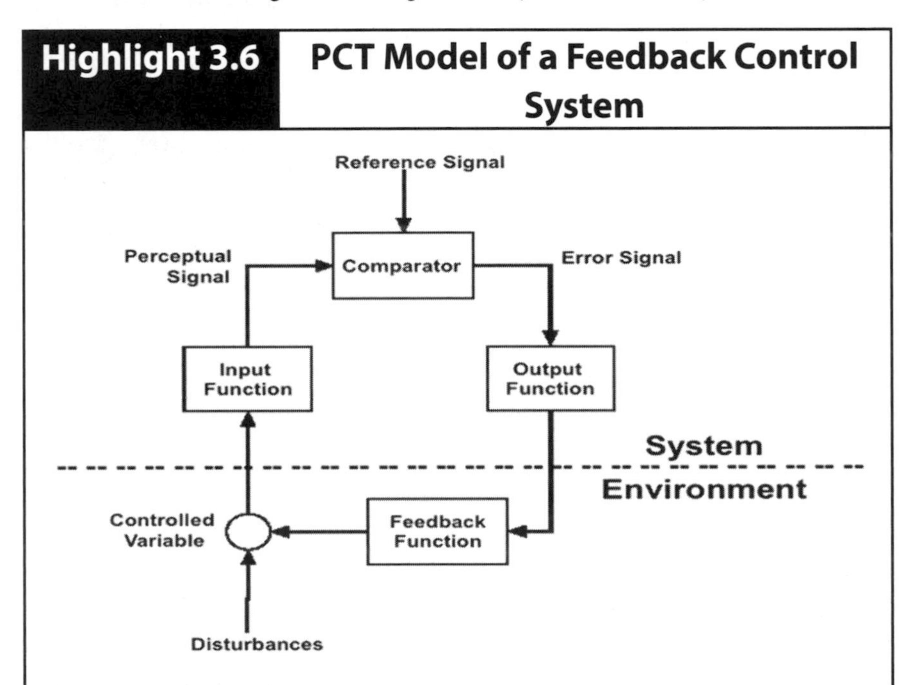

This shows the basic control unit of behavior on which Perceptual Control Theory (PCT) is built. There are many of these systems at the same level and many levels. The environment of higher-level systems includes lower-level systems. The effect of the output adds to the effect of the disturbance on the input.

Sources: Powers, *Living Control Systems III,* 28; Marken, *Doing Research on Purpose,* 63; Carey, *Method of Levels,* 37.

Highlight 3.6 shows Powers' general model. It is worth studying because it helps explain so much about what you do. Given what we have discussed in previous sections, the model's parts should be somewhat familiar to you. As indicated by the Highlight, what you do is based on continuously operating loops of *perceptual signals, reference signals* that reflect your references, *comparators* that produce *error signals* when there is a difference between your perceptions and reference signals, *output functions* that produce behavior, mostly muscle movement, and physical effects on something in the environment that is a target of your behavior.

The target, sometimes called a "controlled variable," is also affected by other influences in the environment called *disturbances*. Your senses serve an *input function* that provides perceptual signals about the state of the target, and you behave to reduce important differences between those perceptual signals and your reference signals. In other words, you do things to control your perceptions.

Highlight 3.7 **Hierarchical Arrangement of Control Systems**

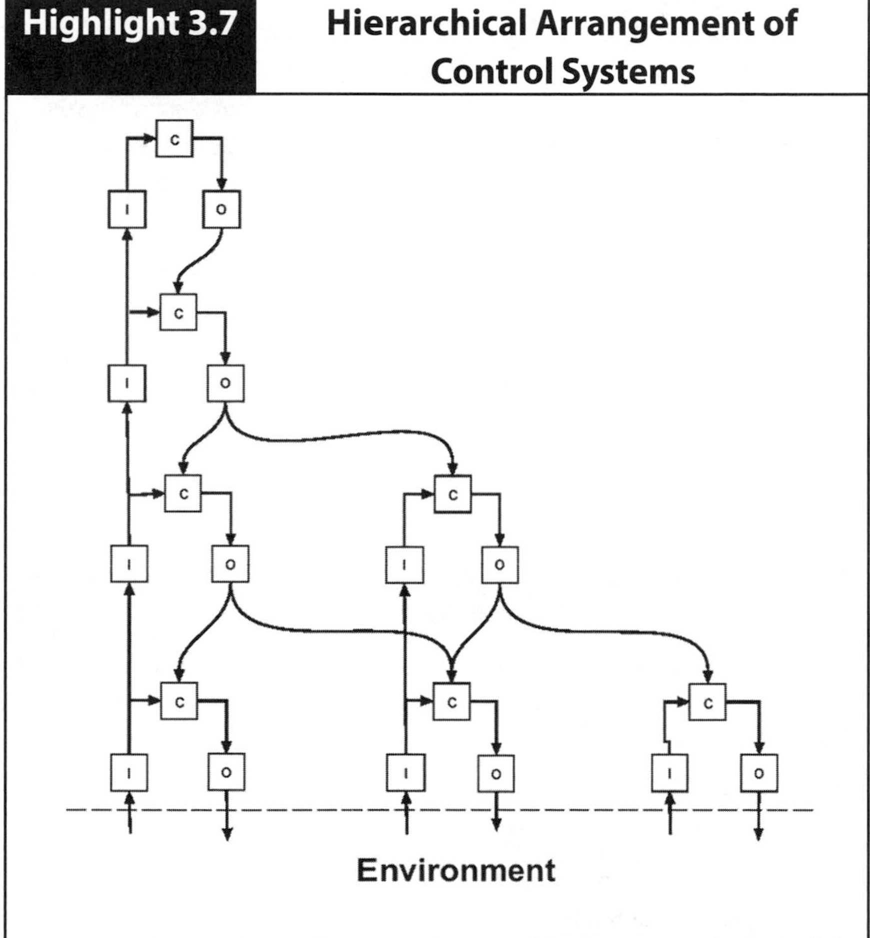

Key: Neural signals flow up from input functions (I) to comparator functions (C) resulting in error signals to output functions (O) that stimulate reference signals down to lower level comparator functions that may result in thinking/planning or output function (O) behavior and action at the system/environment boundary.

Sources: Vancouver, "Self-Regulation," 314; Powers, *Living Control Systems*, 278.

Your body contains hundreds if not thousands of such "control systems"—a concept that comes from engineering and the tradition of cybernetics. Powers estimates, for example, that about 800 such systems control your muscles.[5] Many other control systems are arranged hierarchically, as indicated by Highlight 3.7, and supply reference signals "downwards" to the control systems affecting your muscle movements and glandular secretions. Many of these control systems are linked and operating simultaneously, for instance, controlling your eyes as you read this book, holding your head upright, and regulating your breathing.

As indicated by Highlight 3.7, the hierarchy of control systems consists of different levels of perceptions, references, and outputs. These levels have been tentatively labeled by Powers as follows:

- *System Concepts:* such as personality, Christianity, a family, science, physics, a basketball team, human rights, marriage, the barber shop, and human organizations—the Boy Scouts and General Motors.
- *Principle:* honesty, kindness, loyalty, fairness, justice, sanctity, heuristics, "rules of thumb," and other general strategies.
- *Program:* alternative sequences and possibilities for acting, such as eating out at a fancy restaurant, Chinese buffet, or a fast food place; and playing a game of chess.
- *Sequence:* things that occur in a certain order such as starting a car, singing a melody, and following a cooking recipe.
- *Category:* words denoting similar characteristics of objects and events, such as the color red, apples, fruit, dog, expensive, democracy, and excellence.
- *Relationship:* such as in front of, behind, on, in, above, under, taller, greater than, and causation.
- *Event:* a transition that has a beginning, middle, and an end, such as opening a door, sitting down, manually shifting gears in a car, and the explosion of a firecracker.
- *Transition:* perception without using words of movement or change, such as rising, shrinking, rotating, increasing, decreasing, and turning.
- *Configuration:* perception without using words of patterns of sensations, of objects and things.

- *Sensation:* perception without using words of colors, tastes, sounds, muscle tension, force, and warmth.
- *Intensity:* the perceptual input level responding, for example, to light, air vibrations, pressure, touch, balance, and chemicals in the air. Perceptions at this level are the signals generated by sensory nerve endings. Outputs at this level are muscle movements and glandular secretions such as hormones.[6]

The labels shown in Powers' hierarchy indicate the types of perceptions that occur and are controlled at that level. For example, the sensation level deals with the control of sensations such as those of light, sound, and muscle tension. The sequence level deals with the proper order of doing things, such as baking a cake or checking the oil level of a car. The program level deals with alternative possibilities and making choices to deal with *if... then* situations that arise, such as *if* this is the situation, *then* I will do that. For example, if a guest wants coffee, I will make some and serve it in a coffee cup along with cream, sugar, and a spoon. But if the guest wants juice, I will take the juice from the refrigerator, pour it into a glass, and give it to the person along with a small napkin. Likewise, the principle level deals with the control of perceptions concerning principles such as kindness to others, doing a good deed daily, the Golden Rule, and that of chess players called "control of the center."[7]

As indicated before, each level of control systems provides reference signals to control systems below it as shown by Highlight 3.7. Those lower-level control systems then provide reference signals to control systems at an even lower level, and so on down until the lowest level is reached and you do something such as move your muscles. For example, suppose you are driving to work. As shown by Highlight 3.8, your highest reference for doing so might be a desire to obtain money. As a result, you control to obtain money. That may result in a reference to "go to work," since you know that you obtain money by working, and so you control going to work by driving your car there. Driving your car, in turn, has references such as "keep in the center of the road." That reference affects control systems linked to the muscles in your hands and arms that move the steering wheel so your perception of the car in the center of the road's lane is controlled. In this way, your higher-level references and control systems send reference signals to lower-level control systems. At the same time, perceptual signals from your lower control systems are sent to higher ones so that appropriate adjustments can be made based on error signals that result from differences between your perceptions and references.

Highlight 3.8 Driving to Work: Hierarchy of Control Systems

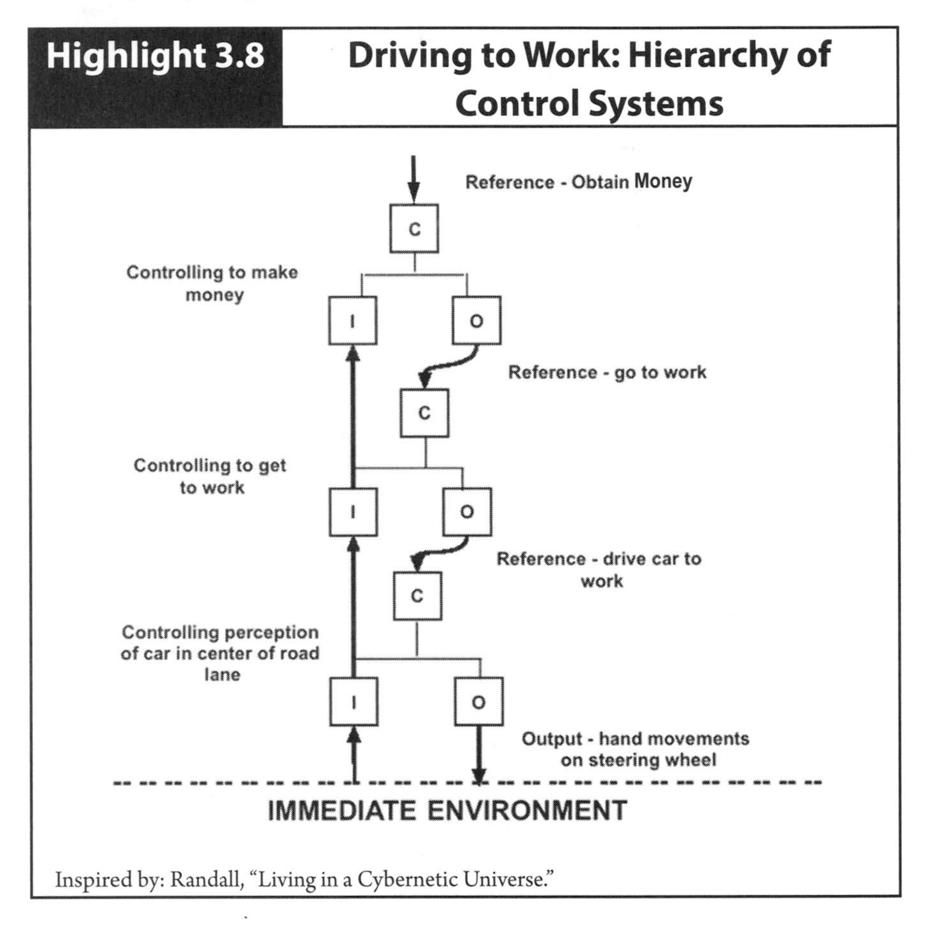

Inspired by: Randall, "Living in a Cybernetic Universe."

To summarize, the PCT model and the processes described by it provide a basic explanation of why you do the things you do—including why the perceptions known to affect behavior, shown in Highlight 3.3, have the effect they do. You do things to control your perceptions so that they match your references and their reference signals. Many of these perceptions, references, and behaviors are learned. However, the most basic control systems, those intrinsic to your survival, are built into you to ensure that, insofar as possible, you live and continue to survive!

A slight variation of Powers' thinking may help clarify relationships between the parts of PCT. That variation is called The Target Model.

The Target Model

The Target Model developed by Fred Nickols is based on PCT. It is easy to understand and clearly shows how different parts of Powers' theory fit together and affect what you and others do.[8] Nickols developed this model, in part, to provide managers with insights into ways of understanding and improving their performance and the performance of those who work for them. However, the model is relevant to what everyone does.

As shown in Highlight 3.9, the references that you have are represented by *Goals* that specify the desired state of a targeted variable. Your perceptions of what is happening outside or inside yourself are represented by *Perceptions*. In particular, this term refers to your perception of the state of your Goals' *Targets*. Your Goals and related Perceptions are compared by you, *The Performer*. If there are differences that are important enough, you do something to reduce or eliminate the gap between your Goals and Perceptions. What you do is represented by your *Actions*, which you carry out to bring the state of your Targets to their desired values.

Highlight 3.9 The Target Model

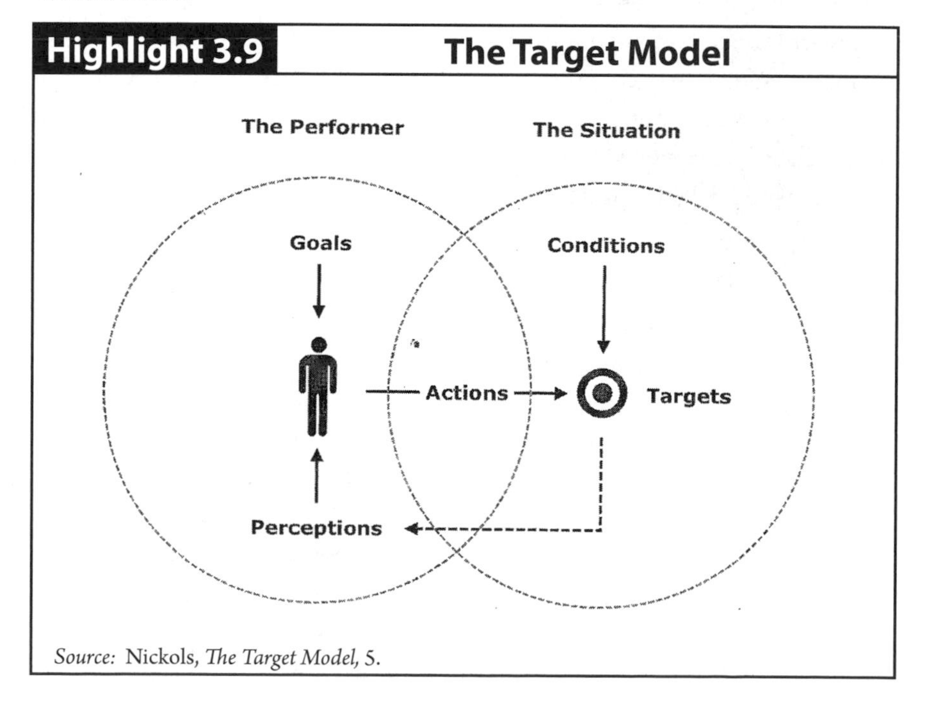

Source: Nickols, *The Target Model*, 5.

Your environment at the time being considered is called *The Situation.* The situation includes the Targets that you want to match your Goals. In addition to your Actions, often other conditions influence your Target. These other conditions range from natural influences, such as the weather, to what other people do. On the diagram, they are shown as *Conditions.* Where your Actions and the Conditions come together is shown by the symbol of a Target. The state of the Target is a result of the combined effects of both your Actions and the situation's Conditions on your Target. In other words, your *Action* is an intervention aimed at affecting the *Target* in order to reduce or eliminate the gap between your *Perception* of the *Target* and your *Goal.*

The dotted line shown on Highlight 3.9, from the Target to your Perception, represents feedback from the Target to your Perception of the Target's state. You may directly obtain feedback and resulting perceptions of the Target such as by looking at it or hearing it, or you may obtain other forms of feedback such as reports or reactions from other people, by making measurements of the Target, or in other ways.

Finally, the two overlapping circles distinguish between you *The Performer* and what is outside you, your external environment or *The Situation* as shown on the diagram. The overlapping part of the circles shows that your *Actions* are common to both you and your environment. That is, you (The Performer) do things that affect your environment (The Situation).

If the model could speak, as Fred Nickols has written, here is what it might say:

> We human beings select or Target certain variables in our environment (T). We set Goals (G) defining the state we want those variables to be in and we compare our Perceptions (P) of the actual or current state of those variables with the goal state we have set. If these two states are not acceptably aligned we engage in Actions (A) aimed at reducing or eliminating any gap between the two. For the most part we are successful. We are "living control systems" and we do a pretty good job of controlling things. However, there are other Conditions (C), other actors and factors that affect the same variables we want to control. For the most part, these other actors and factors pose minor disturbances to our control and we can compensate for them. But on occasion they can overwhelm our best efforts. Our control is far from perfect.[9]

The model basically indicates that if you perceive things to be different from your goals, and the difference is important enough, you try to reduce or eliminate that difference if you think you can successfully do so.[10] In short, *when there is a gap between your perceptions and your goals, you act to eliminate the gap if the goal seems attainable and is important enough to pursue.* If there is no difference or gap, if it's not important, or you don't have the ability to reduce the gap, there is no reason for you to act. Here are some examples:

- Suppose some friends are coming to visit. You look around and see that your house or apartment is a mess (a Perception, P) which when compared to your reference state or Goal (G) of having a tidy place when friends come over results in a gap, or in PCT terms, an error signal. You then intentionally take action (A) to clean up so that your place looks neater (P). You continue tidying up until you are satisfied (P = G) or run out of time (a Condition, C) or energy to do more.
- After eating, food is stuck in your teeth. You use a toothpick or your finger to remove it. Your perception of the condition of your teeth (P) differs from your goal (G) of not having food stuck in your teeth, so you take action (A) until you perceive (P) that the object is removed and no difference exists between your perception and goal (P = G).
- Your boss sets a goal (G) for you to do something related to your work. You do what she said. You take action (A) until you perceive (P) that the goal is achieved (P = G).
- As a salesman, your targeted sales volume is not what it should be; it is too low (P). There is a difference between your perception of sales (the state of your Target, T) and your goal concerning your sales target (G). You then take action (A) and work harder to increase your sales until your perception of sales (P) equals or exceeds your goal (G).

If these examples seem familiar and understandable, it is because such perceptions and behavior occur every day. They are a part of everyday life. Highlight 3.10 shows that such thinking is also the stuff of the extraordinary things that some people do.

Highlight 3.10 **Why Do People Spy?**

Believe it or not, it's mainly because of MICE!

The intelligence community has an acronym for helping it understand why people spy. It is "***MICE***." MICE refers to the references of spies that affect what they do. Here's what the letters stand for:

- ***M*** is for money, to live better, satisfy greed, or deal with debt.
- ***I*** is for ideology, including political opinions, and cultural or religious beliefs.
- ***C*** is for coercion, such as blackmail or threats against one's family.
- ***E*** is for ego and feelings of importance or excitement.

Some people also add an R to form ***MICER***, with the last letter being:

- ***R*** for revenge and grievances toward a workplace, organization, or country.

As you may realize, the intelligence community's thoughts are compatible with PCT and recognition of the importance of a person's references in affecting what they do. People spy mainly to achieve perceptions that match one or more of their reference levels: for money, for their ideological principles, to avoid unpleasant circumstances because of coercive threats, to obtain thrills and boost their egos, and sometimes for revenge.

For more details look at wikipedia.org under the heading "Motives for Spying" and see Herbig and Wiskoff, *Espionage Against the United States.*

Now that you know how your perceptions, references, and behavior are related, let's move on to other aspects of your body that also affect what you do.

Other Internal Systems and Factors

Your Feelings

Linked to your perceptions, references, and ways of doing things are your feelings about those things. Your feelings are often an unrecognized factor that affect a great deal of what you do because they are important to your survival! They are a way of determining, quickly and broadly, what is good for you and what is not. Good feelings are linked to those things and actions that usually help you achieve your references, maintain your homeostatic autopoietic being, and survive. Bad feelings are linked to those things and actions that may interfere with achieving what you desire, disrupt your body's homeostatic states, and act against your survival.

You usually are receptive to and do things that feel good—things for which you have positive perceptions, that seem like the right thing to do, and that usually result in achievement of your references. You also tend to avoid things and avoid doing things that seem bad or give you a bad feeling. For example:

- If you perceive a threat to yourself such as a thief, you will have a negative feeling. Your reference of being in a nonthreatening condition and feeling pleasant will be different from the situation you are facing. Using Powers' terms, an error signal occurs that feels bad. And so (1) your stress response will be automatically activated, and (2) your behavior will be effected to help change your perception. You may freeze to hopefully avoid harm, call for help, defend yourself, run away, or do something else such as use pepper spray on the threatening person.
- If you perceive a pleasant state such as a friend coming to visit, you will do things to ensure or continue that pleasant state such as smile, give the person a hug or a pat on the back, and talk pleasantly.

Your feelings also apply to awareness of your body. You sometimes feel somewhat unpleasant or bad, such as when you feel sick, tired, fatigued, or exhausted. Or you may feel great. Such feelings affect what you do. If you feel sick, perhaps you will rest, avoid strenuous activity, and avoid doing things that you routinely do such as going to work or to a party. If you feel great, happy, or in love, you will probably behave differently.

Highlight 3.11 The Feeling of Pleasure and Pain

"The pleasure-pain mechanism serves to motivate the animal's actions toward those things which its survival requires and away from those things which threaten its survival."

(Binswanger, *Biological Basis of Teleological Concepts,* 133)

"Humans ... make decisions based on their feelings, and these in turn specify what is, or was, in their best biological interest."

(Johnston, *Why We Feel,* 179)

"I believe that the very purpose of our life is to seek happiness."

(Lama and Cutler, *The Art of Happiness,* 13)

In fact, your feelings have such a powerful effect on your behavior that some scholars think that feelings are the ultimate deciding factor behind what one does. Lawrence B. Mohr, for example, has developed an *affect-object paradigm* where in the encounter of affects (i.e., feelings) the strongest always wins. See Highlight 3.12 for details.

Highlight 3.12 Mohr's Affect-Object Paradigm

Lawrence B. Mohr proposed a model that helps explain why you do one thing rather than another. By *affect* he means either an emotion currently felt and stimulated by an object or a residue—not necessarily conscious—of past emotion that he calls an affect tag. By *object* he refers to neural representations of contenders for influence over behavior at the moment including physical objects, concepts, thoughts, behaviors, words, and feelings. He says, "The objects in the affect-object field continually vie with one another by means of their attached affects and the affect tags on associated objects for influence over both attention ... and the motor system" in which "the strongest wins in this struggle for influence" (pp. 72-3). In addition to unconscious influences, conscious reasons and decisions are also capable of influencing behavior by virtue of the affect associated with those events. However, according to Mohr, "*What is determinative... is the encounter of affects, where the strongest always wins*" (p. 95).

The diagram below summarizes some of Mohr's thinking about the relationship of conscious reasons and decisions to behavior, where the affect-object system is ultimately the determining factor.

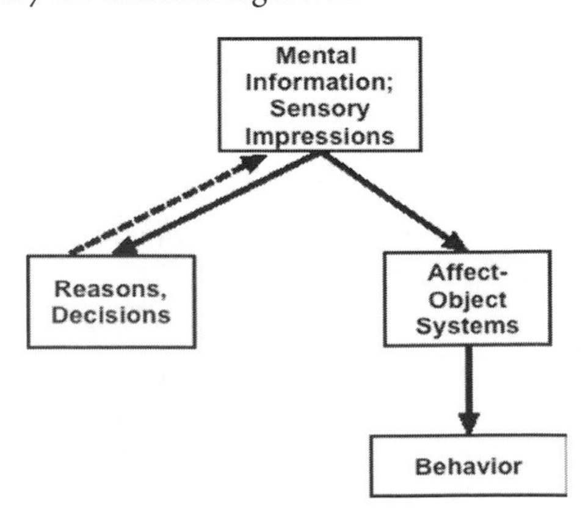

Source: Mohr, *Causes of Human Behavior,* 94.

Similarly, Michael Cabanac has proposed that people choose among doing different things based on the amount of pleasure aroused by the behavior chosen.[11] You do things that give you more pleasure because those feelings are usually linked to behaviors and conditions that are good for you—behaviors that tend to support your autopoietic existence by eliminating or reducing error signals and maintaining your homeostatic states.

To summarize, your feelings are a useful guide that influence what you do. They are automatic, quick-acting survival indicators. However, as guides, your feelings are not infallible. That is why consciousness is important. Conscious thought can help overcome your initial feelings and actions when you think they are inappropriate. But even then, the feelings (i.e., affect) associated with your conscious thought ultimately prevail according to Mohr—a view that I am inclined to accept. At the least though, even if Mohr is not 100 percent correct, your feelings are useful and powerful guides that affect what you do. And if your feelings are intimately linked to "error signals" as indicated earlier, Mohr's views are neatly compatible with those of Powers' Perceptual Control Theory.[12]

Hormones and Your Endocrine System

Your endocrine system and the hormones it produces affect your behavior indirectly. That is, your hormones do not cause you to do specific things. But they may increase your sensitivity to certain conditions or affect how your body functions. For example, "Women who live together for an extended periods of time eventually begin to synchronize their menstrual cycles."[13] Such cycles and synchronization are a hormone effect that indirectly affects behavior. Another example is your fight or flight response. Chapter 2 noted that this response prepares you to take action by increasing your respiration rate, heart rate, blood flow to your muscles, and blood sugar levels as a result of hormones released by your endocrine system. These effects, however, do not determine the specifics of what you actually do.[14]

Hormones also affect your memory and emotions. They may generally sensitize your perceptions and action systems, affect mood in adult women, and influence aggressive behavior, social attachments, and sexual behavior.[15] In addition, they affect the brain's organization during development, particularly sexual differences.[16]

For example, when you are sick, the hormone *Interleukin-1* triggers white blood cells to divide and proliferate. This hormone also makes you sleepy, changes your threshold of pain regulation so that your body aches, makes you run a temperature, decreases your appetite and sex drive, and generally makes you feel bad.[17] Such effects may decrease your activity level, cause you to be less efficient than usual, and result in resting or sleeping more. But the hormone does not cause you to take medicine, or sick leave, or stay at home and rest. It does, however, make such behavior more likely, especially when such behaviors are pre-existing tendencies that you have. The same holds for other hormones, as Highlight 3.13 indicates for the hormone testosterone.

Highlight 3.13 Testosterone and Aggression

"What it [testosterone, a male sex hormone] does is exaggerate pre-existing social tendencies toward aggression. If you want a…metaphor…, testosterone does not turn on some radio playing martial military music. What testosterone does, if and only if the music's already on, it turns up the volume.... Testosterone is exacerbating, amplifying, increasing the volume of what was already there by social experience.... Testosterone is not causing aggression. It amplifies pre-existing tendencies."

(Sapolsky, *Biology and Human Behavior*, Part 2, 156-157)

Your Body's Structure, Condition, and Maturation

Your body limits and affects what you do. Without your muscles, skeleton, nerves, heart, and lungs, you cannot do anything. However, some parts are more directly related to your behavior than others. For example, your muscles and skeleton permit you to move. However, they also limit your movement in some ways. For example, you can't turn only your head around to see what is behind you or jump over tall buildings. What you put into your body such as drugs, and overall conditions such as your general health and fatigue also affect what you do. So do your sense organs, both in terms of the sensations you have and the concepts and general behaviors that you develop. For example, if you had a highly developed sense of smell like that of a dog, you would surely behave differently and probably think differently too.

The maturity of your body also affects what you do. When you were a child, for example, you were not physically capable of doing the same things that you can do as an adult. You did not have the muscles and bone size to do much. Your heart and lungs were also smaller. Your reactions times were longer, and your hand-eye coordination, balance, and judgment of movement were not as good.[18]

Mentally, you were not fully mature. For example, the front part of the brain is not completely developed or myelinated until about the mid-20s. This area (the prefrontal cortex) affects behaviors such as planning, reasoning, decision-making, and control of impulsiveness and related social behavior. Such brain underdevelopment seems to be a reason why juveniles are more reckless and less inhibited in what they do.[19] So, depending on your age, the relative development of your brain may be a factor in your behavior.

An important point is that, *even though many parts and systems of your body affect your behavior, your brain and nervous system are especially important.* They affect almost everything you do—your thoughts, feelings, sensations, perceptions, goals, plans, dreams, and actions. Such effects become especially noticeable when the nervous system is disturbed through injury or the use of drugs. Injury may result in disorientation, memory loss, and paralysis. Similarly, drugs such as aspirin, caffeine, opium, anesthetics, and alcohol may cause effects such as the reduction of pain, increased or decreased alertness, hallucinations, and sleepiness. These then affect perception and behavior. However, besides your brain and nervous system, most effects of your body on your behavior are relatively indirect and not very obvious, except, perhaps, when you are sick.

Your Genes

Although your genes are largely responsible for the structure of your body, they do not really cause you to do things except when expressed in your neural structures as actions such as coughing, sneezing, or crying as an infant after you were first born. Even though the popular media has reported that genes for many complex behaviors have been found—such as for schizophrenia, bed-wetting, and aggression—such claims are bogus misleading statements, sometimes prompted by a few "genetic evangelists" and passed on by journalists and newscasters who lack enough scientific background to know better than to report misleading information.[20]

The effects of genes on behavior are much more indirect. Genes produce proteins, and while those proteins may influence behavior, many other factors are involved. For example, Highlight 3.14 indicates that if we consider an antisocial act of aggression, various factors are important including the situation, opportunity for action, emotional provocation, and a quick assessment of whether the act is worth doing or not. In short, *genes do not cause you to behave in specific ways.*[21]

Highlight 3.14 Some Factors That May Lead to an Antisocial Act

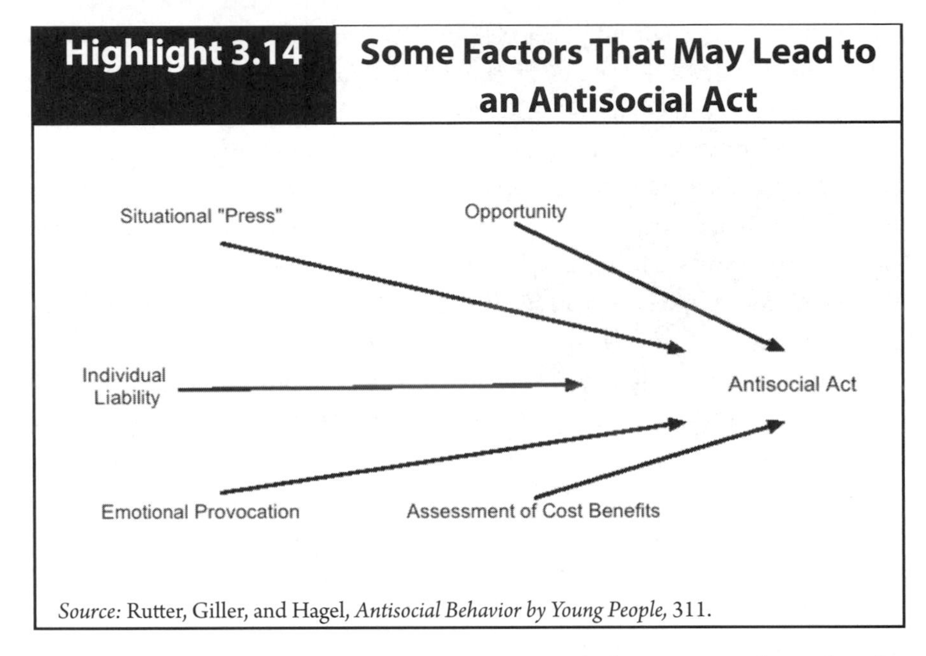

Source: Rutter, Giller, and Hagel, *Antisocial Behavior by Young People,* 311.

However, genes do indirectly influence your behavior. As indicated in the previous section, your genes are responsible for the general structure of your body—which allows you to do certain things but also limits what you can do. Genes are also assumed to influence temperament evident from early childhood, including your general activity level, emotional responsiveness, distractibility, and willingness to explore.[22] Such tendencies may indirectly influence behavior, but they do not cause you or others to behave in specific ways. In short, genes are structuring causes, not direct triggering causes of behavior as explained more in Highlight 3.15.

Highlight 3.15 Structuring and Triggering Causes

According to Fred Dretske in his book *Explaining Behavior,* a *triggering cause* induces the process of behavior to occur now. It can be viewed as the most direct and immediate cause of a behavior. *Structuring causes* establish conditions that result in the specific behavior that occurs.

For example, for a dog conditioned to salivate at the sound of a bell, the triggering cause of salivation is the bell. Structuring causes include the dog's past experience with food and bells and the effects of that experience on the structure of the dog's nervous system, the structure of which changes slightly as a result of experience. The dog's genes can also be regarded as structuring causes, but more distant ones, since they helped generate the dog's overall nervous system and salivary glands as the dog developed.

For you, an electric alarm clock going off in the morning may be the triggering cause to wake up. Structuring causes for waking up may include setting the alarm the night before, the reason you set the alarm for that particular time, and the electricity supply that was available to activate the clock and its alarm. Your genes are a much more distant, less immediate, structuring cause.

Preview of the Next Chapter

Now that you are aware of major influences that your body has on your behavior, you may be interested to know more about how you became the way you are. Your behavior has been influenced and shaped by many factors, perhaps chief among them being the effect of your social environment upon development of your neural system from the time you were a baby until now. Chapter 4 focuses on these factors and influences.

Further Reading

For information about preferred states:

- Harry Binswanger, *The Biological Basis of Teleological Concepts*. (Los Angeles: Ann Rand Institute Press, 1990). A basic book on human purposeful action.

For information about Perceptual Control Theory (PCT), including the importance of your perceptions and references:

- William T. Powers, *Behavior: The Control of Perception*, 2nd ed. (New Canaan, CT: Benchmark Publications, 2005). The basic PCT textbook.
- William T. Powers, *Making Sense of Behavior: The Meaning of Control* (New Canaan, CT: Benchmark Publications, 2004). An explanation in nontechnical language written for general readers.
- William T. Powers, *Living Control Systems III: The Fact of Control* (New Canaan, CT: Benchmark Publications, 2008). Includes computer demonstrations that can help you better understand PCT.

- Philip J. Runkel, *People as Living Things: The Psychology of Perceptual Control* (Hayward, CA: Living Control Systems Publishing, 2005). A comprehensive introduction to PCT and human functioning.

For information about the Target Model, which is based on PCT but aimed at helping practicing managers, go to Fred Nickols' website at www.nickols.us/articles.html and download publications such as:

- "A Control Theory View of Human Performance in the Workplace," "Helping People Hit Their Performance Targets," "Manage Your Own Performance: No One Else Can," and "The Target Model: A Mainly Visual Presentation."

For information about feelings and behavior:

- Victor S. Johnston, *Why We Feel: The Science of Human Emotions* (n.p.: Helix Books, 1999). An evolutionary functionalist explanation of feelings and consciousness.
- Antonio Damasio, *Looking for Spinoza: Joy, Sorrow, and the Feeling Brain* (Orlando: Harcourt, 2003). A progress report on the nature and significance of feelings.

For information about Mohr's affect-object paradigm:

- Lawrence B. Mohr, *The Causes of Human Behavior: Implications for Theory and Method in the Social Sciences* (Ann Arbor: University of Michigan Press, 1996). Written for the practicing empirical researcher.

For information about endocrinology and behavior:

- Randy J. Nelson, *An Introduction to Behavioral Endocrinology,* 4th ed. (Sunderland, MA: Sinauer Associates, 2011). Provides a good overview of the relations between hormones and behavior, especially in nonhumans.

For overviews of how genes and behavior are related:

- Michael Rutter, *Genes and Behavior: Nature-Nurture Interplay Explained* (Malden, MA: Blackwell Publishing, 2006).
- Matt Ridley, *Nature via Nurture: Genes, Experience, and What Makes Us Human* (New York: HarperCollins, 2003).

For information about triggering and structuring causes:

- Fred Dretske, *Explaining Behavior: Reasons in a World of Causes* (Cambridge, MA: MIT Press, 1988).

Endnotes

1. When you notice or are aware of your goals, objectives, plans, targets, desires, and other references, you are a self-observer. Such self-awareness is often called "consciousness." As you do most things, though, you are often not aware of your referents.

2. These are things that Abraham Maslow referred to at the bottom of his hierarchy of needs, for example in *Motivation and Personality*.

3. Baumeister and Sommer, "Consciousness, Free Choice, and Automaticity."

4. For example, see Marken, "You Say You Had a Revolution."

5. Powers, *Behavior: The Control of Perception*, 93.

6. Descriptions of these levels are contained in Runkel, *People as Living Systems*, 193-214, and Powers, "A Cybernetic Model," 190-208. Personality as a systems concept is from M. Powers, "Control Theory," 60.

7. Runkel, *People as Living Systems*, 208-209.

8. Nickols, *A Control Theory View*. Publications about the Target Model, also called the GAP-ACT Model, can be obtained online by going to www.nickols.us.

9. Nickols, *Tools for Knowledge Workers*. Downloaded 6 July 2013 from http://www.nickols.us/PCT101Primer.pdf

10. Albert Bandura has written a book related to the last part of this sentence, about how people's beliefs in their ability to successfully take action affect what they do. He writes, "Such beliefs influence the course of action people choose to pursue, how much effort they put forth in given endeavors, how long they will persevere in the face of obstacles and failures, their resilience to adversity, whether their thought patterns are self-hindering or self-aiding, how much stress and depression they experience in coping with taxing environmental demands, and the level of accomplishments they realize." From Bandura, *Self-Efficacy*, 3.

11. Cabanac, "Pleasure: The Common Currency."

12. Powers indicates that such a relationship between error signals and feelings does exist. For details, see Chapter 17 of his book *Behavior: The Control of Perception*, 2nd ed., 252-261.

13. Nelson, *Behavioral Endocrinology*, 3rd ed., 358, based on research by McClintok, "Menstrual Synchrony and Suppression," 244-245.

14. An exception to this may be "rape induced paralysis" that apparently affects about 40 percent of rape victims at least to some extent. Such immobility results from

a flood of hormones that impair the brain's thought and emotional processes, according to Sherpa, "Fight, Flight, or Freeze." This reaction seems similar to that of playing dead when attacked, a successful survival strategy of some animals as described by Herzog, "Playing Dead, as a Defense Against Angry Bears and Rapists" from http://www.psychologytoday.com/blog/animals-and-us/201007/playing-dead-defense-against-angry-bears-and-rapists.

15. Hampson, "Sex Differences," 616.

16. Nelson, *Behavioral Endocrinology,* 3rd ed., 51 and Chapter 3.

17. Sapolsky, *Biology and Human Behavior,* Part 1, 148.

18. Berger, *The Developing Person,* 273 & 347.

19. National Institute of Mental Health, *The Teen Brain*; Peper and Dahl, "The Teenage Brain."

20. For example, see "How the Media Promote the Public Misunderstanding of Science," in Goldacre, *Bad Science,* 224-241; also, Rutter, *Genes and Behavior,* 11-12.

21. Rutter, *Genes and Behavior,* 71-72.

22. VandenBos, *APA Dictionary of Psychology,* 928.

CHAPTER 4

HOW YOU BECAME THE WAY YOU ARE

The Big Picture

As indicated before, you are a survivor because of the way your body is organized and because what you do helps sustain your life or, at least, doesn't kill you immediately. Some processes for surviving are built into you, for example, the way your body is organized for breathing, coughing, sneezing, and jerking away from touching something painful, such as a flame. Your general temperament, such as tendencies toward introversion and extroversion, also seems to be built into you. However, much of what you do is learned. From the time you were a fetus inside your mother's body until now, you have been learning—partly because of the way your nervous system works and partly because of your experiences with the world around you. As a result of your ability to learn, you are a flexible being. Without that ability, you would operate more like an insect or a worm, guided only by your reflexes and instincts. Without an ability to learn, you would not be able to do new things or adapt to new situations. Without your ability to learn, you would be either an automaton or dead.

However, most of your learning and resulting behavior has not occurred by chance; rather, most of what you do has been influenced by your environment. You speak the language of the people you grew up with. You behave in stores, libraries, and other places, and wear clothes similar to those around you. Why? Because your environment has socialized you to behave in those ways. If you do not behave or act appropriately as judged by other people, you will often be corrected or punished. Informally, you may be hollered at, ridiculed, frowned upon, ignored, or ostracized. Formally, you may be regulated by the legal, medical, and religious systems of your environment—such as by being arrested, tried, and fined or imprisoned; by being required to undergo counseling, or committed to a mental institution; and, eventually, by facing the afterlife and its effects after your death such as by ending up in the everlasting pain and agonies of Hell (a belief that has shaped the lives of many people towards more socially acceptable ways of behaving). However, if you behave in ways approved by those around you, you will tend to have a more pleasant life. People will be nicer to you. They may smile. They may do favors for you and, as a result, help to increase the quality of your life and your chances of survival.

The Story of Your Life

The way you became the way you are began when the egg and sperm of your parents united. As a result of the genes they provided and the environment in your mother's womb, you developed into a healthy fetus, at least healthy enough to live. When you were born, your bodily processes, structures, and reflexes enabled you to survive.[1] Your breathing control systems ensured an adequate oxygen supply. Your rooting and sucking systems helped you obtain the food and energy you needed to function and develop. Your intestinal and urinary systems permitted you to eliminate waste products. When you were uncomfortable, you cried—an action that usually obtained a helpful response from your caretakers. Your nervous, circulatory, digestive, and other systems worked well enough for you to live.

You also began to learn. Your nervous system functioned in a way that led to perceptions of different events becoming associated with what you did. Your crying became linked to being handled, to different spoken sounds, to food, warmth, and a change in visual scenery that occurred when you were picked up. Random movements you made, such as smiling and moving your arms and legs, sometimes led to different events—perhaps another change in visual scenery such as your mother's smiling face, or touching something that felt different and possibly moved or made a sound, such as a colorful mobile object hanging above you. Such novelty and events were pleasurable and over time you learned to repeat the movements and sounds that led to experiencing them.

As your body developed you also did other things randomly. Some of those movements became linked to perceptions that were pleasant or unpleasant. You learned to repeat doing things that led to pleasurable outcomes and avoid doing things that led to unpleasant events. Some of those outcomes or consequences occurred naturally—such as the novelty of a changing visual field when you walked and the pain of touching something hot. Some consequences were social, such as a parent smiling and talking to you softly when you did something nice, or shouting at you when you did something that he or she didn't want you to do. So you learned to do things that met with other people's approval and avoided doing things that led to their disapproval.

You were a great imitator. You mimicked the sounds of people talking. You did what other people did. In fact, you learned a great deal by imitating other people and you still do in your dress, conscious actions, and other behavior. You also developed attitudes towards people, things, and events either as a result of direct experience with them, by watching and listening, or in other ways such as by reading.

Other people also helped you to learn by teaching you. You learned new words and behaviors such as how to brush your teeth and tie your shoelaces. You learned how to put on your clothes and eat properly. You went to school and learned many things there. You also learned from friends, from watching television, and from people where you worked.

At first you were attentive and concentrated when you did new things, such as buttoning a shirt, riding a bicycle, or driving a car. After many repetitions, your behavior often became fairly automatic. Now, you often behave without thinking about what you are doing.

Even as a grown-up, people around you continue to affect what you do. You continue to behave in ways that lead to other people's approval and avoid doing that which leads to disapproval and other unpleasant consequences. You greet people when they greet you. You behave normally when you are in public places such as stores, movie theaters, and on streets. You behave within the limits of the law when a police officer is around. You do these things, unless you have good reason not to.

In short, you have learned to live in ways that are as pleasant as possible within the environment that surrounds you. You have become the way you are in part because of your genetic heritage and in part because of your environment, your experiences in that environment, and what you have learned to do and not to do.

Let's look more closely at some of these factors and processes that have influenced you and what you do.

Your Genetic Heritage

Your body's structure depends largely on the genes that you inherited from your parents. That is why you look more like them than you look like a dog or a horse. Your environment has also affected your body to some extent as indicated by Highlight 4.1. For example, the structure and organization of your nervous system has been affected by the experiences you have had over your lifetime—and still is. By affecting your body, such influences also affect what you do, at least to some extent.

Highlight 4.1 **Examples of Environmental Effects on Body Structures and Functions**

Environmental Agent	Effect
Stimulation as an infant; visual and auditory experience; physical contact and other interactions with people, objects, places, and events; experiences with the external world from birth to death	More complex neural circuits and other changes in neural circuitry that are related to intelligence, visual, auditory, and other neural development. Infants thrive on touch, physical contact, and varied experiences, especially during the first months of life. Adults are also affected.
Nutrition	Malnutrition in utero and during infancy often results in smaller brains, substantial IQ deficits, slower language development, and difficulties with fine motor skills. Starvation reduces immunity from germs. Height increases with good nutrition.
Prenatal and postnatal infections: rubella (German measles), genital herpes, syphilis, chicken pox, cytomegalovirus (a common virus), and toxoplasmosis (a common parasitic disease).	Possible hearing loss and deafness in babies; cataracts and resulting childhood blindness or decreased binocular vision; possible malformations of the brain that lead to mental retardation.

Alcohol consumption by mother during pregnancy	Child mental retardation and malformations at high doses; cognitive delay at moderate doses; possible hyperactivity or speech deficits if mother was alcoholic. Fetal alcohol syndrome is the leading cause of mental retardation in the United States.
Smoking by mother during pregnancy	Low birth weight and behavioral deficits of children.
Mother's elevated body temperature during pregnancy	Neural defects of children.
Narcotic use by mother	Lower birth weight of child. After birth, poorer maternal care could affect child stimulation, care, and bonding with resulting effects including poor attention span, child's lack of guilt after misbehaving, impulsiveness, poor coordination, hyperactivity, irritability, and speech problems.
High air temperature (heat stress)	Sweating and, over time, the production of dilute sweat, plasma volume expansion, increased heart rate, and increase in body temperature during exercise, all of which help the body to lose heat. Effects can include reduced reaction times, fatigue, confusion, dizziness, headache, aggression, and fainting.
High altitude	Hyperventilation; increase in number of red blood cells; increase in blood flow to the brain and heart to maintain oxygen supply; altered hormone secretion; and fatigue.
Participation in war and contact sports; community and domestic violence; accidents	Concussions and other traumatic brain injury; effects may include problems with attention, concentration, memory, anxiety disorders, insomnia, aggression, irritability, vision impairment, headache, dizziness, fatigue, depression, apathy, personality change, and psychosis.

Note: Some effects shown are definite, such as neural changes due to stimulation and experience; other effects often occur but not always.

Sources: Eliot, *What's Going On;* Piantadosi, *The Biology of Human Survival;* McCrea, *Mild Traumatic Brain Injury;* Grilly and Salamone, *Drugs, Brain, and Behavior;* Vasterling, Bryant and Keane, *PTSD and Mild Traumatic Brain Injury;* McLaughlin and others, "Causal Effects."

However, you may recall from Chapter 3 that genes have only an indirect influence on behavior. As Highlight 4.2 indicates, your genes do not cause you to do anything specific. But they, and the structures they help produce, do interact with your environment and affect what you do. In fact, your body has been interacting with the environment around you ever since you were a fetus in your mother's womb. And this contact and interaction has profoundly influenced you and what you do.

Highlight 4.2 **The Effects of Genes**

"We are almost never justified in using the language or concept of a 'gene for' a behavior or psychiatric trait. The relationship between genes and behavior is too contingent and indirect for such language to be appropriate."
(Flint, Greenspan, and Kendler, *How Genes Influence Behavior,* 212)

"Genetic instructions are of great importance to our natures, but they are not destiny."
(Ehrlich, *Human Natures,* 9)

"Genetic material does not produce finished traits but rather interacts with environmental experience in determining developmental outcomes."
(Bronfenbrenner and Ceci, "Nature-Nurture Reconceptualized," 571)

"The fact that we speak English has nothing to do with genetics, but the fact that we can talk does."
(Gallagher, *The Power of Place,* 218)

For example, did you know that some of your food preferences may be affected by what your mother ate when you were an embryo? Such preferences are the result of an interaction between you and your maternal environment.[2] Other environmental influences are more evident. The language you speak and think with was learned through interaction with people around you. How you dress is also based on your environment: if you are a bushman, you may wear skins; if an American, underwear, shirts or blouses, pants or skirts; if a Hindu Brahman woman in India, perhaps a sari, at least some of the time. Also, most of the routine normal behavior you exhibit in public, including how you act in behavior settings, roles you carry out, and manners, are a result of how you have been socialized by your parents, teachers, peers, and others. Such socialization effects are a result of your ability to learn. Let's take a closer look at that ability, because it has greatly affected your perceptions, references, and resulting behavior.

Highlight 4.3 — Some Learning-Related Definitions

Learning *n.* the acquisition of novel information, behaviors, or abilities after practice, observation, or other experiences, as evidenced by change in behavior, knowledge, or brain function.

(VandenBos, *APA Dictionary*, 2nd ed., 594)

Reorganization: The process of changing the forms of functions in the hierarchy of control systems.

(Powers, *Behavior: The Control of Perception, 287*)

Why Learning is So Important

Your ability to learn greatly improves your chances for survival. Learning helps you to cope with your changing environment rather than to do things in fixed, unvarying ways. You do the things you do mainly because you have learned to do them that way. You are such a good survivor in part because you have learned ways to survive. You have learned to avoid things that may harm you and to do things that help you to live. Learning helps you to maintain your autopoietic being.

Without the ability to learn, your behavior would be mostly fixed, like that of an insect. You would only be capable of fixed reflexive behavior. For example, when you were a baby you could do only a few things: cry when uncomfortable, suckle, root for a nipple, sleep, urinate, defecate, and kick and move your arms randomly. Now, because of your ability to learn, you can probably talk and read, brush your teeth and keep yourself clean, buy clothes and dress yourself, use a television and computer, drive a car, and perhaps earn a living.

With your ability to learn, your actions are flexible and adaptable to new situations and experience. With your ability to learn, you also benefit from experiences of others by talking with them, observing what they do, reading books and articles that they have written, and by applying their lessons at home, school, and work.

Some Principles of Learning

The Law of Effect

When you read "The Story of Your Life" earlier in this chapter, you saw that you learn to do things that lead to pleasant outcomes and avoid doing things that have unpleasant outcomes. This relationship, called the Law of Effect, was described by the psychologist Edward Thorndike more than 100 years ago. Thorndike wrote, "The Law of Effect is that: *Of several responses made to the same situation, those which are accompanied or closely followed by satisfaction to the animal will, other things being equal, be more firmly connected with the situation, so that, when it recurs, they will be more likely to recur; those which are accompanied or closely followed by discomfort to the animal will, other things being equal, have their connections with that situation weakened, so that, when it recurs, they will be less likely to occur. The greater the satisfaction or discomfort, the greater the strengthening or weakening of the bond.*"[3]

William Powers, the perceptual control theorist, suggests that such feelings of satisfaction and discomfort result from whether or not your reference values are being achieved. Unpleasant, unsatisfying feelings result when there is a difference between what you perceive and your reference value for that perception: the greater the difference, the greater your negative feeling. For example, the more your body needs water, the more unpleasant is your feeling of thirst. Alternatively, pleasant feelings result when your reference values are being met and when related error signals are reduced or eliminated. For example, you tend to feel good after achieving a goal you have been working toward, such as obtaining a good grade at school, being hired for a job that you wanted, eating a nice meal after being hungry, or drinking water when thirsty.

In other words, considering the Law of Effect using Powers' ideas, you do things that reduce error signals and avoid doing things that lead to greater error signals.[4] More formally, The Law of Effect can be stated as shown in Highlight 4.4.

Highlight 4.4 **The Law of Effect from a PCT Perspective**

Behavior leading to consequences that reduce or eliminate error signals is more likely to occur again in similar situations perceived by an organism. Conversely, behavior that produces error signals or fails to reduce such signals is less likely to occur again in similar situations.

Reorganization of Your Control Systems

According to William Powers, you learn to do things that reduce differences between your perceptions and reference values. If what you are doing does not reduce important differences or does not reduce them fast enough, you will do something else until the difference is reduced or eliminated. If you have no basis in experience for dealing with the difference and do not know what to do, you may act randomly until by chance you happen to do something that works. When this happens, especially after several successes, you learn to do whatever you did to reduce the error signals that were occurring.

Powers calls such learning "reorganization," since the control systems involved are changed or reorganized in the process.[5] This reorganization seems to correspond to changes in your neural networks. The control systems of your nervous system are reorganized to repeat those behaviors that help to reduce error and achieve the consequences that you desire or need to survive.[6] By doing so, the reorganization system responds to error and functions in the same way as other homeostatic processes within your body.[7]

Highlight 4.5	**Reorganization**

"Reorganization changes the way errors are handled...."

(Vancouver, "Depth of History," 330)

"Reorganization alters the properties of the control systems involved in behavior."

(Powers, *Behavior: The Control of Perception, 2nd ed.*, 188)

Association

Related to the process of reorganization is your body's tendency to identify patterns and linkages between different events in your environment. For example, you tend to notice things that happen at the same time or that follow one after the other. They become linked within your brain, so that when one occurs, you either automatically react or more consciously assume that the other will also occur. For example, if you eat a certain food and become ill, you may associate that food with your sickness and avoid eating it again. Also, during pleasant and unpleasant experiences, your feelings become associated with the experience.

That is one way that you develop the likes and dislikes you have. For example, if a dog bit you, you may be wary of dogs. If, instead, you had positive experiences with dogs, you probably like them.[8] Similarly, if you did well in math at school, you may have a liking for math or, if you did poorly, you may dislike it. If you enjoyed someone's company, you may like that person; if someone hurt you or was unpleasant towards you, you may not like that person and you may even dislike people who are similar to them. In such cases, you have formed a linkage between people, objects, or events, and your feelings about them.

Advertisers try to develop such associations. Examples are many and include perfume and car ads that link their products to sex appeal; cigarette ads that associate their brands with relaxation, good times, and rugged masculinity; and McDonald's "happy meals" that are linked to clowns, pleasant images, and toys. In this way, perfumes, cars, deodorants, shampoo, prescription drugs, and even financial services are often associated with appealing features such as lovely sunsets, beautiful bodies, and smiling faces. Logos, too, have been linked to athletic contests and winning athletes, as Nike has done.[9]

Richard Perloff, a scholar of persuasive techniques, writes: "Association is perhaps the most important reason why advertising succeeds."[10] Advertisers try to develop associations so people feel favorably towards their names and products. By doing so, some have undoubtedly shaped some of your attitudes toward what they sell and causes that they promote.

Even simple exposure to neutral people and objects has been shown to result in feelings of liking those things. We seem to develop comfortable feelings towards non-threatening things that we frequently observe or hear. This happens especially if we have not developed strong feelings toward them before. Advertisers and politicians recognize this phenomena and try to influence how you feel about their products or them. According to Richard Perloff, "Mere exposure is a strong, robust persuasion phenomenon" and repetition at first "is a positive experience, reducing uncertainty, inducing calm, and bringing on a certain amount of pleasure."[11] However, Perloff warns that, after a certain point, repetition can lead to negative feelings toward an ad or product. According to him, "This is one reason why companies like McDonald's and Coke [Coca-Cola] frequently switch slogans and change advertising agencies. They want to prevent wear-out and preserve the effect of a novel campaign slogan."[12]

Sometimes, however, a tendency to notice relationships and regularities leads to false conclusions. That is, you may perceive relationships that do not really exist or associate objects, events, and experiences that do not regularly occur or occur randomly, by chance. Such false, invalid, and sometimes dysfunctional associations are the basis of many superstitions, prejudices, stereotypes, addictive habits, fears, and post-traumatic stress disorder (PTSD, as discussed in Highlight 4.6). For example, superstitious rituals are sometimes established this way and then transmitted as a part of the culture where they developed.[13] In China, for instance, the word for "eight" sounds similar to the word which means "prosper" or "wealth," and there is resemblance between the two digits "88" and a popular decorative design composed of two stylized characters meaning "joy" or "happiness." Thus the number "8" is considered lucky by many Chinese, such that the opening ceremony of the 2008 Summer Olympics in Beijing began on 8/8/08 at 8 seconds and 8 minutes past 8 p.m. local time, and a telephone number with all digits being eights was sold for US$270,723 in Chendu, China.[14] Similarly, in the United States, some people believe that possession of a rabbit's foot and hanging a horseshoe in one's home bring good luck, whereas a black cat running across one's path is a sign of bad luck. Do any superstitions affect your behavior? If so, it's natural. But you might recognize them for what they probably are—false associations!

Highlight 4.6 Post-Traumatic Stress Disorder

"I was raped when I was 25 years old. For a long time, I spoke about the rape as though it was something that happened to someone else. I was very aware that it had happened to me, but there was just no feeling. Then I started having flashbacks. They kind of came over me like a splash of water. I would be terrified. Suddenly I was reliving the rape. Every instant was startling. I wasn't aware of anything around me. I was in a bubble, just kind of floating. And it was scary. Having a flashback can wring you out. The rape happened the week before Thanksgiving, and I can't believe the anxiety and fear I feel every year around the anniversary date. It's as though I've seen a werewolf. I can't relax, can't sleep, don't want to be with anyone. I wonder whether I'll ever be free of this terrible problem."

Post-traumatic stress disorder (PTSD) develops after a terrifying ordeal that involved physical harm or the threat of physical harm. The person who develops PTSD may have been the one who was harmed, the harm may have happened to a loved one, or the person may have witnessed a harmful event that happened to loved ones or strangers.

PTSD was first brought to public attention in relation to war veterans, but it can result from a variety of traumatic incidents such as mugging, rape, torture, being kidnapped or held captive, child abuse, automobile accidents, train wrecks, plane crashes, bombings, or natural disasters such as floods and earthquakes.

People with PTSD may startle easily, become emotionally numb, have trouble feeling affectionate, be irritable, become more aggressive, and even become violent. They avoid situations that remind them of the original incident, and anniversaries of the incident are often very difficult.

Most people with PTSD repeatedly relive the trauma in their thoughts during the day and in nightmares when they are asleep. These are called flashbacks. Flashbacks may consist of images, sounds, smells, or feelings and are often triggered by ordinary occurrences, such as a door slamming or a car backfiring on the street. A person having a flashback may lose touch with reality and believe that the traumatic event is happening all over again.

Not every traumatized person develops full-blown or even minor PTSD. Symptoms usually begin within three months of the incident, but occasionally emerge years later. Some people recover within six months, while others' symptoms last much longer. In some people, the condition becomes chronic. PTSD affects about 7.7 million American adults.

Certain kinds of medication and psychotherapy can often effectively treat the symptoms of PTSD.

Source: National Institute of Mental Health, *Anxiety Disorders*.

To summarize, you routinely and automatically establish associations between things and events that you sense, including between your perceptions and feelings. These associations are learned and affect the way you think and act. Some associations are learned naturally. Some are a result of trauma. Some are taught. Others are formed and sometimes manipulated by people such as advertisers, politicians, and even con artists. These associations, when formed within you, affect how you perceive things and what you do.

Neural Networks and Learning

Let us now focus on your neurons and nervous system. As you know, your behavior depends on the structure and organization of your nervous system. However, you may not know that the organization of your nervous system has changed over your lifetime.

When you were born, you had many more neurons in your brain than you do now. Those that you used as an infant and child continued to exist. Those that were not used disintegrated and disappeared—a clear example of the folk saying, "Use it or lose it." In addition, neurons that were used formed connections with other neurons and with continued use over time, those connections were strengthened. Such connections are the basis of your control systems. They are visibly expressed by your actions.

The process of establishing connections between neurons, as you experience and learn new things, continues to this day. As William James, psychologist and philosopher, wrote in 1890: "The entire nervous system is nothing but a system of paths between a sensory *terminus a quo* and a muscular, glandular, or other *terminus ad quem*. A path once traversed by a nerve-current might be expected to follow the law of most paths that we know, and to be scooped out and made more permeable than before; and this ought to be repeated with each new passage of the current.... So nothing is easier than to imagine how, when a current once has traversed a path, it should traverse it more readily still a second time."[15]

Although James recognized this possibility long ago, the details have been discovered only recently. Researchers have found that the structure of neural networks really is affected by our experiences. The changes that occur result in what we call learning—your neurons and neural networks are changed, reorganized, and restructured to some extent. (See Highlight 4.7 for details.)

Highlight 4.7 Neural Processes of Learning and Memory

It is clear that short-term and long-term forms of learning involve changes in neural circuits—that is, among the ways that neurons are organized and connected among themselves and with the muscles and glands that they activate. The changes are based upon chemical and electrical changes within individual neurons and the release and reception of substances called neurotransmitters in the spaces between neurons. When short-term learning occurs, the chemical and electrical changes are fairly temporary. Learning that occurs over longer periods of time involves not only such changes, but also the development of new physical connections between concerned neurons or between the neurons and the muscles or glands which they stimulate. Alternatively, the strength of some connections may be reduced or connections eliminated entirely leading to memory loss.

In any case, most learning and memory is a result of connections distributed within networks of neurons. Periodic reactivation of neural relationships and connections that exist may be necessary to maintain them and the associated learning and memory that results.

The diagrams below give an idea of what happens at the gaps called synapses between two neurons (A and B) during short-term and long-term learning. As is shown, during short-term learning, small packets of neurotransmitters are released into the synapse by A to establish a connection between the neurons. When longer-term learning occurs, one or more additional synapses are established between these neurons.

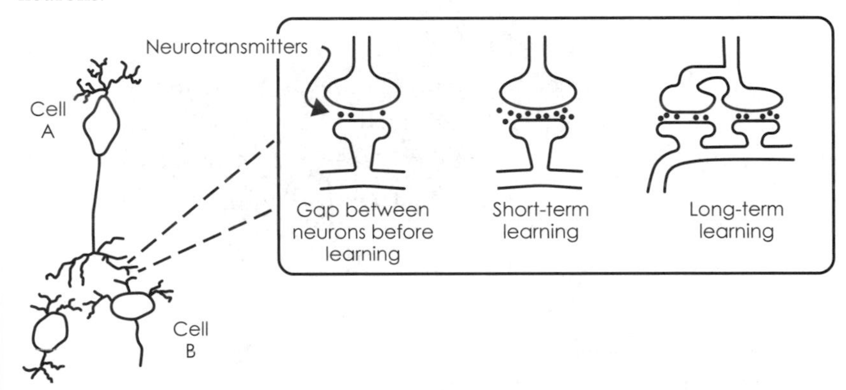

Given that an average neuron forms and receives 1,000 to 10,000 synaptic connections with other neurons, muscles, and glands, and that the human brain has about 85 billion neurons, these diagrams illustrate only a small, but important, part of what happens when learning occurs.

Sources: Kandel, *In Search of Memory*; Kandel et al., *Principles of Neural Science*, 4th & 5th eds.; Squire et al., *Fundamental Neuroscience*.

How You Learn: Observable Aspects

Most of what you do is a result of what you have learned to do. So, let's look at how you learn and have learned in the past.

Trial and Error

As mentioned before, you were born with some simple reflexes such as sucking and crying. You also were able to randomly move your arms, legs, head, and eyes, make sounds, smile, and do a few other things. In the process of thrashing around, your random actions caused certain things to happen. For example, as you waved an arm, perhaps it touched a bright toy hanging above you and caused it to move. Doing this randomly over time, your nervous system began to associate the two happenings by forming neural connections and reorganizing your nervous system, and you eventually learned to move your arm in such a way as to touch and move the toy in a more controlled manner. In similar ways, you learned to grasp things and make sounds that resulted in a response from your mother or other adults, for example, by smiling more or saying "Mama." Your random behaviors became associated with certain perceived consequences. So you learned to do certain things in order to obtain pleasant perceptions and avoid unpleasant perceptions.[16]

As you grew older, you continued to learn the same way. If something you did didn't result in a desired perception, you did something else either randomly or after some thought. You tried doing different things until you either obtained what you wanted (a desired perception) or gave up and did something else. In this way, through trial and error, you learned how to do many things and developed many skills. You continue to do so.

Observation and Imitation

Another important way that you learned was by observing and imitating what other people did. As a child, you were a fine imitator. You observed and imitated actions such as facial expressions, sounds that others made including language, and other behaviors of people around you. You still do so today.

Imitation has been called the "social glue" that makes us successful social animals. We imitate others a great deal—their postures, speech, and facial expressions—usually unconsciously. Doing so often leads to greater rapport and liking, to smoother interactions, and even to higher income.[17] You imitate clothing styles, polite and cooperative behavior, and generally conform to what others are doing. In novel situations when you don't know what to do, you often imitate others and learn what to do. For example, at a formal dinner, you may look to see what fork or spoon others are using before you begin to eat an unfamiliar serving. Or, arriving by airplane at an unfamiliar airport, you may see what direction your fellow passengers are going and then follow them. The same holds for other people too. Nearly everyone imitates a lot.

In fact, we benefit a great deal by observing and learning from others around us. We get an idea of how to do things by watching what others do. Observing others saves time that might otherwise be wasted by trial and error. Needless mistakes may be avoided. When we learn to perform a new role, the opportunity to watch someone who is successful and embodies what we want to become is invaluable.[18] That is one reason why apprenticeship training is so successful: an apprentice is able to see a master in action.

Many attitudes are also formed by observing others and seeing the results of their behavior. At the workplace, when we observe a coworker do something, such as use a new technique that has good results, we may try to do likewise. If we observe that our boss and other workers often come to work late, we may start to come late. Similarly, performance standards, attitudes toward safety procedures, and work ethics are often learned by observing others. The same holds at home, school, and other behavior settings such as at stores, playgrounds, theatres, and places of worship. For example, many of your attitudes have been shaped by seeing what your parents and other caretakers did.

Besides directly observing others, you have also observed and learned by seeing others on media such as television, movies, and videos. Some of your values, beliefs, and behaviors have been learned this way, haven't they? For example, if you are fashionable, you surely have been affected by what you've observed as being fashionable.

Highlight 4.8 **Observational Learning**

"Most human behavior is learned by observation through modeling."

"The people with whom one regularly associates ... delimit the behavioral patterns that will be repeatedly observed and, hence, learned most thoroughly."

(Bandura, *Social Foundations of Thought*, 47, 55)

Instruction by Others

You also learned a lot from people who taught you to do things. Your parents or others taught you to brush your teeth, tie your shoelaces, eat properly, and perhaps ride a bicycle. You have spent literally thousands of hours being instructed by adults including teachers at school who taught you how to read, write, and play sports. At the workplace too, you may have gone through orientation and induction programs, been shown how to accomplish tasks by your bosses and other workers, attended training sessions, and perhaps even been mentored.

Others have also tried to teach, persuade, and influence you in different ways, ranging from door-to-door salespersons and cause advocates, to producers and writers of television programs, advertisements, newspaper and magazine articles, and e-mail. Such techniques have obviously affected what you know, what you do, and how you feel about things—although you may not be aware of the extent of those influences. For example, though you may have had no direct experience with them, you probably know something about political parties, terrorists, space exploration, child abuse, global warming, North Korea, Africa, and Antarctica.

In short, you've learned many things as a result of other people trying to influence what you know and do.

Self-instruction

In addition, you have sometimes actively learned things on your own. You may have done this by reading, talking with other people, trying to see how others do what you are interested in, trying things out, and practicing a skill. Hobbies, a need to repair things, problems at work, and other interests and needs you have had, may have resulted in self-initiated learning activities. Also, if you have tried to change your behavior in the past, some of your efforts may have involved self-instruction.

Experience and Learning by Doing

Another important way you learn, one that overlaps with other methods, occurs when you simply do things. That is, you learn by doing—as you have learned throughout your life. When you do things by trial and error, you are learning by doing. When you successfully imitate others, you are learning by doing. When you self-instruct yourself and take resultant action, you are learning by doing. For example, when you learned to ride a bicycle or drive a car, you learned to do those things by doing them. When you learned to tie your shoes, you learned by tying them. The same holds for the languages you speak and how you cook, read, write, use a computer, and do your job (if you have one). You learned to do these and other things by doing them.

The old folk sayings, "Experience is the best teacher" and "Practice makes perfect," express the essence of learning by doing—although some scholars point out that feedback and perception of the results of one's actions are also needed. That is, if you don't know the consequences of your actions, if you don't know how you did, you usually won't improve what you do.

Think about the workplace. Most jobs require that people actually do them in order to become proficient. For example, many skills that senior managers use when they manage are learned and refined through experience—first as beginning supervisors and then in more complex jobs. Similarly, nurses become more competent over time because they experience a variety of patients, cases, and situations. They know from experience what to do and when to do it. The same holds for good workers in any field. That's one reason experienced workers are often paid more. Expertise is largely a result of experience and related learning by doing.

At home too you have learned by doing. Do you have children? If so, you know what I mean. Are you married or have a partner? If so, haven't you learned about one another and things to do and avoid doing while living together? Do you cook? If so, you undoubtedly learned by actually cooking and experiencing the results of your efforts. Similarly, we learn a lot in unique ways by direct experience with the world around us and by doing things, as indicated by Mark Twain's quotation shown in Highlight 4.9.

Highlight 4.9	Experience and Learning
"A man who carries a cat by the tail learns something he can learn in no other way." (Mark Twain)	

Social and Other Environmental Influences

Socialization and "Social Control"

Socialization is defined as "the learning of a culture" and "the process by which individuals acquire social skills, beliefs, values, and behaviors necessary to function effectively in society or in a particular group."[19] Most of what you do is a result of how you have been socialized by the people and society around you. Your language, manners, and most normal behaviors that you exhibit are a result of socialization and the influence of others.

At home, your parents were important socialization agents. As you spent more time away from home, your peers became important influences. At school, teachers performed important socializing roles. At work, your bosses and co-workers did. Other socializing mechanisms include the mass media, religion, and government.

As a baby, child, and adolescent, your parents helped you learn by praising and punishing you, by doing things as simple as smiling when you did acceptable things and frowning or shouting when you did unacceptable things. Perhaps they gave you a candy or let you watch television if you did certain things and were a "good boy" or a "good girl"—that is, when you behaved as they wanted you to behave. Or, they may have punished you or told you something such as that Santa Claus would not give you any presents at Christmas if you were bad.

As you became older, others began to exert influence—your relatives, friends, teachers, and bosses. Others in the community did too. If you acted normally as expected in public, other people dealt with you in predictable ways: such as by acting towards you pleasantly; by thanking you when you did something to help them; and by letting you be a part of their groups, teams, and clubs. But if you acted abnormally or behaved in ways that displeased others, you often experienced less pleasant behavior. This may have included stares, frowns, corrections, harsh talk, and perhaps even social isolation—unpleasant things that made you feel bad and you tried to avoid. Such influence continues today.

More formal attempts to influence you and others also exist. The police and mental health specialists can be called upon when your behavior exceeds normal bounds and is not quelled by those around you. If you are religious, your behavior may also be affected, in part, by the prospect of eternal pleasure or eternal damnation that awaits you in heaven or hell.

In other words, you have been molded to do things that result in pleasant or neutral responses from others, and avoid doing things that result in unpleasant responses from other people and perhaps ultimately from God. See Highlight 4.10 for details. The techniques of influence shown in that highlight will be discussed more in later chapters about your environment and changing your behavior. But for now, you should realize that your social environment has had a tremendous influence on you and the things you do.

Highlight 4.10 "Social Control" Techniques

"Social control" refers to techniques used by members of society as they attempt to assure the norm-conforming behavior of others. Examples include:

Informal – a pat on the back, a smile, a kiss, a hug, praise, a compliment, giving recognition of achievement, expression of disappointment or anger from a parent or others, corrections, frowning, glaring or staring at someone, verbal insults, scolding, silent treatment, withdrawal of friendship, ignoring, shunning, avoiding, excluding, rejecting, expulsion, time-outs, an assault, beating, expulsion, ostracism

Organizational – orientation and training, supervisor verbal feedback, performance appraisals, pay raises, bonuses, public recognition (such as employee of the month awards), incentive programs, employee assistance programs, 360-degree feedback instruments, standard operating procedure manuals, protocols, reprimands, demotion, transfer, termination

Medical – drug or alcohol rehabilitation programs, campaigns to reduce smoking, psychological examination, counseling, therapy, psychoanalysis, drug treatments (for ADHD, depression, and other mental disorders), admittance to a mental hospital

Legal – laws; penalties provided by the state for violations of criminal, administrative, or civil law, including restraining orders, Anti-Social Behaviour Orders (in the UK and Ireland), use of force by the police, fines, home confinement, mandated community service, imprisonment, exile, and death; also rewards, such as tax breaks and deductions

Religious and Supernatural – commandments, sermons, Sunday school, religious publications, love bombing,[a] excommunication, the concept of sin and its consequences, rewards and punishments individuals receive upon their death

[a] "love bombing" refers to showering new recruits with extra love and attention.

Sources: Chriss, *Social Control;* and Williams, *Ostracism.*

Note: The term "social control" is in quotation marks since whether someone can actually control another person is open to question.

Your Physical Environment

As indicated before, your physical environment has also affected how you became the way you are and what you do. This environment includes gravity; the air around you, including its temperature, humidity, and pressure; the weather; the surface you are on; landscaping; structures around you; chemicals you were exposed to before and after your birth; germs and other microbes; objects, including books, computers, furniture, machines, communication devices, eating utensils such as chopsticks and forks; and whether you grew up in a city, town, farm, natural setting, or other location. Highlights 4.1 and 4.11 provide examples of such effects.

Some things in your physical environment are resources used to achieve your goals and other desired perceptions. Others provide opportunities for you to act. Some directly affect the biology of your body and your survival, both positively and negatively.

Highlight 4.11 Lead Exposure: Another Example of Physical Environmental Effects

The metal lead has been shown to have a number of effects on the human body and behavior. No safe threshold for lead exposure apparently exists. Young children are especially susceptible to lead poisoning since their growing bodies absorb more than adult bodies.

Among other effects, exposure to lead interferes with development of the nervous system. Signs and symptoms in children include irritability, learning disabilities, and behavioral problems such as mental retardation, hyperactivity, poorer attention, lower vocabulary, poorer eye-hand coordination, longer reaction times, decreased arithmetic and reading skills, decreased fine motor skills, and antisocial behavior such as aggression. In adults, depression and anxiety occur as does irritability, poorer attention spans, headaches, muscular tremors, forgetfulness, loss of memory, hallucinations, delirium, malaise, impotence, dizziness, greater levels of conflict in interpersonal relationships, and poorer verbal reasoning, reaction times, hand dexterity, and posture balance.

Most exposure is by ingestion, such as of lead paint by children, and by breathing lead-contaminated dust and (in some countries) lead-gasoline exhaust fumes. Most lead exposure is preventable.

Major Source: Agency for Toxic Substances and Disease Registry, "Lead: Health Effects."

We'll discuss your environment more in later chapters. For now, it's enough to realize that both your physical and social environment affect (a) your learning, including your knowledge, skills, and attitudes, (b) your body's structure and organization, and (c) what you do.

Preview of the Next Chapter

Now that you are aware of how you became the way you are, you may be interested in knowing more about what you actually do. For example, how often do you do things automatically, without really thinking? How regular and routine is your behavior? When is your behavior more unique? The next chapter deals with these matters. The contents may surprise you.

Further Reading

For information about interactions among genes, behavior, and the environment:

- Matt Ridley, *Nature via Nurture: Genes, Experience, and What Makes Us Human* (New York: HarperCollins Publishers, 2003). Discusses the idea that genetic effects are vulnerable to experience.

For information about childhood development:

- Lise Eliot, *What's Going On In There?: How the Brain and Mind Develop in the First Five Years of Life* (New York: Bantam Books, 1999). An easily readable book.
- Hetty Vanderijt and Frans Plooij, *The Wonder Weeks: Eight Predictable Age-Linked Leaps in Your Baby's Mental Development* (Arnheim, Netherlands: Kiddy World Promotions, 2008). A book whose authors found their findings explained well by Perceptual Control Theory.

To learn more about the neural basis of memory and learning:

- Larry R. Squire and Eric R. Kandel, *Memory: From Mind to Molecules* (New York: Henry Holt and Company, 2000). Easily readable for a technical book.

For scholarly information about imitation and observational learning:

- Susan Hurley and Nick Chater, eds., *Perspectives on Imitation: From Neuroscience to Social Science,* 2 vols. (Cambridge, MA: MIT Press, 2005).
- Albert Bandura, *Social Foundations of Thought and Action: A Social Cognitive Theory* (Upper Saddle River: Prentice Hall, 1986). See Chapter 2, "Observational Learning".

For a readable account of "social control":

- James J. Chriss, *Social Control: An Introduction* (Cambridge, UK: Polity Press, 2007).

For information about influences of the physical environment on child development, a scholarly article is:

- Gary W. Evans, "Child Development and the Physical Environment," *Annual Review of Psychology* 57 (2006): 423-451.

Endnotes

1. It seems that what many people call "reflexes" are the observable behavior of control systems. For details, see Chapter 7 of Powers, *Behavior: The Control of Perception.*

2. Eliot, *What's Going On In There?*, 176-77, 191-92.

3. Thorndike, *Animal Intelligence*, "Chapter V: Laws and Hypotheses for Behavior Laws of Behavior in General," 1911, accessed May 5, 2016 from Classics in the History of Psychology, http://psychclassics.yorku.ca/Thorndike/Animal/chap5.htm.

4. We can also restate Powers' ideas using concepts mentioned in Chapters 1 and 2. That is, you do things that maintain or restore homeostatic balances that are essential to maintaining your autopoietic life, and avoid doing things that create homeostatic imbalances that can lead to your death.

5. The idea of reorganization as expressed by Powers was suggested by W. Ross Ashby in his notion of "ultrastability," and independently by Donald T. Campbell as "blind variation" and "selective retention" according to Powers, "An Outline of Control Theory," 27-28.

6. Other scholars have noted somewhat similar ideas that seem to overlap with Powers' thinking concerning the relation between error signals and learning. For example, Jarvis in *Toward a Comprehensive Theory* uses the term "disjuncture" to express the unease that exists when we cannot cope automatically with our situation. Such times when we do not automatically know what to do or how to respond, he calls a "teachable moment." Similarly, along with Glaser in his article "The Reemergence of Learning Theory," D. Ford and Learner recognize that "learning results from practice that *reduces and minimizes error* (i.e., discrepancies)" in their book *Developmental Systems Theory*, 170. Referring to Skinner's idea of selection by consequences, Ford and Learner state that "a positive reinforcing consequence is one that helps reduce the discrepancy between current and preferred states" (183)—a mode of thought similar to that of Powers who indicated that although people do learn by reinforcement, the process is not understood by psychologists and others who lack an understanding of control theory. In Powers' words, "[People] learn by varying what they do until their actions cause something to happen that reduces intrinsic error. Some people call that reinforcement, but they don't understand how it really works" (Powers, in a posting on November 26, 2007, to csgnet@listserve.uici.edu).

7. L. Morris, Mansell, and McEvoy, "The Take Control Course," 7.

8. Of course, you can also develop attitudes toward things and events through observation and imitation. For example, if you observed that your mother was afraid of dogs, or spiders, or snakes, you may have learned to be afraid of them too.

9. Perloff, *The Dynamics of Persuasion*, 284-86.

10. Ibid, 285.

11. Ibid, 282-83.

12. Ibid, 283.

13. Skinner, *Science and Human Behavior*, 86-87.

14. "Numbers in Chinese Culture," accessed October 27, 2009 from Wikipedia, http://en.wikipedia.org/wiki/Numbers_in_Chinese_culture.

15. James, *The Principles of Psychology*, vol. 1, 108-109.

16. Fascinating demonstrations of how this happens are contained in Chapter 8 of Powers, *Living Control Systems III.*

17. Dijksterhuis, "Why We Are Social Animals," 208.

18. Douglas A. Bernstein et al., *Psychology*, 2nd ed., 286; Manz and Sims, "Vicarious Learning," 106; Kram, *Mentoring at Work*, 113.

19. Chriss, *Social Control*, 20; VandenBos, *APA Dictionary*, 2nd ed., 997.

CHAPTER 5

HOW REGULAR AND ROUTINE IS YOUR BEHAVIOR?

The Big Picture

What you and others do is predictable in many ways—especially when your references are known! Mostly, you do things regularly and routinely without thinking about what you are doing. You behave automatically most of the time. Why? Because doing so results in consequences that you desire or need to survive. In fact, the consequences that result are even more regular than the behavior that produced them. As the models in Chapter 3 indicate, the reasons behind your behavior are not hard to understand. But, how regular and automatic your behavior is may surprise you.

Focusing on your daily life, most of what you do is done fairly routinely with little or no thought. For example, your routine observable behaviors probably include brushing your teeth; your sleeping habits; and most of what you do at home, school, and work. When you go to the bathroom, you almost certainly defecate in the toilet and not elsewhere in the room—certainly not elsewhere where you live unless you are disabled by illness. When you go to an expensive restaurant, you also do things in regular ways, such as waiting to be seated, looking at the menu, ordering drinks and food, eating, paying, and probably leaving a tip. And the same goes for most of your other behavior: when you play games according to the rules; go to church or temple and sit quietly, listen, and sing or respond at appropriate times; and when you walk.

However, there are times when you do things very differently and often more consciously. These are mainly (1) when what you are doing is not producing the results that you want, (2) when you are facing a new, novel situation for which your usual behavior is not working to produce perceptions that you desire, and (3) when you are learning to do something new.

How Regular is What You Do?

The Extent of Your Routine Behavior

Although you make slight adjustments depending on what you perceive, you generally do the same basic things to perceive consequences that you desire. Why? Because, what you routinely do is usually effective in producing the results that you want.

The same holds for other people. In fact, the regularity of behavior and the references that drive that behavior are reflected in a number of words that describe what people prefer and do. Words such as habits, personality, skills, normal behavior, norms, roles, customs, traditions, and culture all indicate regularities that have been noticed by others. See Highlight 5.1 for details.

Highlight 5.1 Words That Reflect Regularities of Behavior and Related References

Habit: "a well-learned behavior or automatic sequence of behaviors that is relatively situation specific and over time has become motorically reflexive and independent of motivational or cognitive influence—that is, it is performed with little or no conscious intent" (VandenBos, *APA Dictionary*, 2nd ed., 479).

Routine: "a customary or regular course of procedure" or "commonplace tasks, chores, or duties as must be done regularly or at specified intervals" (Maddux, "Habit , Health and Happiness," 337, quoting *Random House Dictionary*).

Personality: "the enduring configuration of characteristics and behavior that comprises an individual's unique adjustment to life, including major traits, interests, drives, values, self-concept, abilities, and emotional patterns" (VandenBos, *APA Dictionary*, 2nd ed., 782). For example, a tense person, a talker, a liberal, a team player, dependable, honest, stress-free, reserved, an introvert (Howard, *Owner's Manual*, 137).

Skill: "an ability or proficiency... Motor skills are characterized by the ability to perform a complex movement or serial behavior quickly, smoothly, and precisely. Skills in other learned tasks include basic skills, communication skills, and social skills" (VandenBos, *APA Dictionary*, 857).

Normal: "Conforming, adhering to, or constituting a typical or usual standard, pattern, level, or type" (*Webster's II*, 803).

Norm: "a standard or range of values that represents the typical performance of a group or of an individual ...." (VandenBos, *APA Dictionary*); "Norms are ideas about what is correct behavior for members of a particular group" (Triandis, *Culture and Social Behavior*, 100).

Role: "a coherent set of behaviors expected of an individual in a specific position within a group or social setting" (VandenBos, *APA Dictionary*, 804). For example, the roles of sales clerk, shopper, policeman, priest, doctor, teacher, manager, mother.

Custom: "a traditional behavior, ritual, or action that is transmitted through the generations and is defined by a culture as appropriate or desirable for a particular situation" (VandenBos, *APA Dictionary*, 253).

Culture: "1. the distinctive customs, values, beliefs, knowledge, art, and language of a society or community. 2. the characteristic attitudes and behaviors of a particular group within society, such as a profession, social class, or age group" (VandenBos, *APA Dictionary*, 250).

In addition, most of your routine behavior is usually carried out unconsciously, with little or no thought. Although precise estimates are lacking, John Bargh, a psychologist at Yale University, suggests that more than 99 percent of your actions are like this.[1] From simple and common behaviors such as walking, eating, and brushing your teeth to more complex behaviors such as reading this book or driving a car, you do these things without thinking much about what you are doing. You do them for the most part automatically and in the same general way at different times.

But one important point before we go on: Regular or routine behavior means generally similar patterns of behavior, not behavior that is exactly the same from one time to another. Your regular and routine behavior is not exactly the same from moment to moment and at different times, in part, because you and your environment are not exactly the same over time. So, you adjust what you do. For example, although you may drive to work regularly, your muscles do not move exactly the same way each time, because of the actions of other cars, the effects of wind pushing your car, road conditions, traffic lights, and pedestrians. In other words, your control systems vary your behavior to suit what you perceive from moment to moment so that your perceptions more closely match your references.

Similarly, although you routinely eat, brush your teeth, and put on clothes every day, the positions and movements of your hand, arms, head, and body are not exactly the same each time. The goals and other references may be similar, but your behavior varies a little in reaching those goals, references, and desired consequences. And if your references change, such as wishing to look nice when going to a wedding, rather than wearing something appropriate to do housework, the clothes you wear will differ but your overall behavior of wearing clothes will be the same. And so, the terms "regular and routine behavior" as used here refer to generally similar patterns of behavior, not to behavior that is exactly the same.[2] See Highlight 5.2 for another example.

The following sections discuss these ideas and your routine, everyday behavior in more detail.[3]

Highlight 5.2 **Walking: An Example of Behavior that is Routine but Necessarily Flexible**

"To take current conditions into account while guiding action toward the intended goal, each movement must be performed a little bit differently. Even highly practiced actions such as walking cannot be a series of rote repetitions with each step exactly like the last because the everyday environment is not like a big gymnasium, with uniform open ground. In real life, paths are cluttered and ground surfaces are infinitely variable. Walking cannot be choreographed or prescribed by a preexisting plan because the everyday environment is not like a fixed obstacle course with all the challenges known ahead of time."

Source: Adolph et al., "Flexibility," 399-400.

Reflexes and Homeostatic Processes

Reflexes are automatic, unlearned, and relatively fixed behaviors that do not require conscious effort to carry them out.[4] They include activities such as coughing; sneezing; swallowing; blinking; itching; withdrawing your body from a painful object; and the rooting, sucking, and crying of babies. Similarly, basic homeostatic processes such as those related to the control of oxygen intake and bodily temperature also result in routine, often unconscious behaviors such as breathing, sweating, and shivering. Such processes, like reflexes, are also usually automatic although they are sometimes consciously controlled.

Besides these reflexes and homeostatic processes, you do many other things regularly every day or at regular times of the year—things that you have learned to do. These include events ranging from brushing your teeth to celebrating birthdays and holidays. You also do things fairly regularly when you are in certain places, such as at stores, libraries, and movie theaters. Let's now consider some of these other kinds of regular behaviors.

Everyday Behaviors

Most of the things that you do during the day and night are done regularly and routinely. You probably sleep in the same place at about the same time most days, in the same way with or without pajamas or a sleeping gown. When you get up, you probably have a fairly stable routine of toilet activities such as going to the bathroom, brushing your teeth, and doing so with the same hand. Other routines may include eating or not eating breakfast and, when you eat, doing so with a relatively fixed variety of foods; traveling to and from school or work about the same way, by bus, car, bicycle, or walking; and, once at school or work, sitting in particular classrooms at about the same places during certain times or doing certain job tasks at work.

Likewise, if you are an American man in a public restroom, you probably don't select a urinal immediately next to another man, unless you have no choice. You don't loiter, wait, or lounge around the restroom. If someone else is there watching you, you also probably wash your hands before leaving.[5]

When you meet people you greet them in typical ways. Your conversations with others also tend to be done fairly automatically, usually without much thought. Other common routine behaviors of many people include watching TV, listening to music, using the internet, reading the newspaper, talking on the phone, texting, showering, washing hands, putting on makeup, doing laundry, and washing dishes.[6] Participating in family rituals such as birthdays, funerals, Thanksgiving, Christmas, Easter, Passover, Sunday dinner, and family reunions is also common—at least where I live.[7]

In fact, if you and others did not behave fairly routinely and predictably in most situations, society would not function well. We would not be able to interact and work together very smoothly. And you would be viewed as a strange, abnormal person.

Highlight 5.3 **Regular Patterns of Behavior**

"Most of the time what we do is what we do most of the time.
Sometimes we do something new."

(Townsend and Bever, *Sentence Comprehension*, 2)

"Daily life is characterized by repetition. People repeat actions as they fulfill everyday responsibilities at work and at home, interact with others, and entertain themselves. Many everyday activities not only are performed frequently but also are performed in *stable circumstances*--meaning in particular locations, at specific times, in particular moods, and with or without interaction partners."

(Wood, Tam, and Witt, "Changing Circumstances," 918)

"The functioning of societies depends upon the presence of patterns for reciprocal behavior between individuals or groups of individuals."

(Linton, *Study of Man*, 113)

"The efficiency and effectiveness of organisations hinges on workers (role incumbents) exhibiting dependable role behavior."

(Stone-Romero, Stone, and Sales, "The Influence of Culture," 331)

"In ordinary, daily life, we find extremely regular patterns of behavior, so regularly repeated that we are astonished and even angry on those rare occasions when they fail."

(Runkel, *Casting Nets*, 153)

"Even when engaged in creative processes, such as writing papers, speaking spontaneously, or driving to novel destinations, the component acts and movements are routinized responses."

(Kirsch and Lynn, "Automaticity in Clinical Psychology," 508)

Behavior Settings

Roger Barker, a psychologist, and his coworkers have determined that our behavior is so routine in many situations that they compare it to the regular patterns of water that occur in a stream or river. That is, during the day, we typically spend our time in a limited number of what they call "behavior settings," such as in one's home, on streets and sidewalks, at school or at a workplace, and in stores. Sometimes we spend time in other behavior settings such as at a church or temple, a sports event, or a restaurant. See Highlight 5.4 for more examples. Barker states, "A person is characteristically on the go throughout his waking hours from one behavior setting to another—from bed to breakfast to school to a ball game and so on through a lifetime."[8]

Highlight 5.4 Examples of Behavior Settings

(Where behavior is routine and recurrent)

Sidewalks and roads
Post offices
Supermarkets/grocery stores
Pharmacies/drug stores
Hardware, appliance, furniture stores
Banks
Restaurants
Taverns/bars
Movie theaters
Nursing homes
Child care centers
School classes
School assemblies
School playgrounds
Sports contests
Circuses and carnivals
Parades
Church worship services
Sunday School classes
Church group meetings
Choir practice
Weddings
Funeral services
Birthday parties
Teachers association meetings
Boy Scout troop meetings
Business meetings
Barber and beauty shops
Motor vehicle service stations
Factories
Rest rooms
Doctor's offices
Libraries
Hotels
Courts
Election sites

Source: Barker and Wright, *Midwest and Its Children.*

In each of these settings, we behave in fairly routine ways. For example, in a church you sit quietly; listen to the priest, pastor, or other religious figure; and pray and sing when appropriate. If you are a spectator at a basketball game, you will sit, occasionally stand, and probably yell and cheer sometimes. If you are a player at the game, you play by the rules. At a library, you tend to talk quietly if at all, look at books, and read. At a store, you look around for things to buy, go to a cashier and pay for things before you leave—that is, unless you are a thief. Routine ways of behaving occur also when you are at work or school, when traveling on roads or sidewalks, and when you are at other places. In such settings, Barker summarizes what you and others do: "People conform in a high degree to the standing patterns of the behavior settings they inhabit.... When an individual's behavior deviates from the pattern of a setting, it is usually symptomatic of mental or physical illness, or of the normal incapacities of extreme youth and age. People, en masse, are remarkably compliant to the forces of behavior settings . . . ."[9]

If you are a normal person, you too tend to behave in routine, fairly predictable ways in the behavior settings you enter each day. If you don't act normally, other people may try to correct what you are doing, remove you, or have you removed from the setting. For example, if you are talking too much with someone else in a movie theater, others around you may ask you to be quiet. If you keep up your disruptive, abnormal behavior, they may call the manager to deal with you. Corrective action of this kind also occurs in other behavior settings to help ensure that you and others behave in normal, expected, and relatively routine ways.[10] These ways correspond to the references that people have for the behavior setting that you are in. At a movie theater and at a library, a common reference is to be quiet. At an upscale restaurant, the usual script or reference program is to wait to be seated, look at the menu, order food from a waiter or waitress, eat the food ordered, receive and pay the bill, and then leave. When driving on public roads, references include stopping at red traffic lights and driving in the proper lane—for example, on the right side of the road in the United States and on the left side in the United Kingdom.

Normal Behavior, Social Norms, and Roles

In behavior settings, such as those shown in Highlight 5.4, your behavior and that of others is fairly predictable. What you and others do is usually normal behavior for that setting. In each setting, you expect other people to behave normally too. That is, you and others have certain expectations or norms (references) for what people should do in each setting. For example, at a store, you expect a cashier at the checkout counter to look at the things you want to buy, tabulate the bill, take payment, and give you a receipt and change, if any. If the cashier spends time talking with a friend and ignoring you, you may become upset because the person is not behaving the way you think he or she should. Similarly, the cashier expects you to behave normally and not leave without paying for things that you are taking. In other words, given your roles as customer and cashier, you both expect the other to behave in certain ways.

Highlight 5.5 Examples of Common Norms that Influence Behavior

Moral: help others in distress; don't hurt others; don't steal; love thy neighbor as thyself.

Reciprocity: return favors done to us by others; do unto others as they have done unto you.

Medical Ethics: do no harm; treat more serious cases first; each patient should be given the fullest possible treatment.

Codes of Honor: do not lie, steal, or cheat; keep your promises; be truthful under oath; in some cultures, seek revenge for an offense.

Fairness: equal sharing; turn taking; equal pay for equal work; equal justice for all.

Etiquette: manners of dress; table manners; greetings; wedding arrangements; behavior at a funeral.

Adults and Children: do not hurt children; children should respect adults; respect your parents.

Work: don't live off other people—earn your income through work; don't be a rate-buster; follow work instructions given by your boss; fairness/equity of procedures and treatment.

Conventional: on which side of the road to drive.

Other: tipping for service; standing in line/queuing.

Sources: Scales et al., *Other People's Kids*; Elster, *Cement of Society* and *Explaining Social Behavior*; Kelley and others, *An Atlas*; Katz and Kahn, *Social Psychology of Organizations*.

Using the ideas introduced in Chapter 3, we can say that you have references or valued norms about how you should behave, and you behave accordingly so your perceptions match those references. Similarly, you have norms and expectations for how others, such as a cashier, should behave. If you perceive that the other person does so, there is no problem. If not, an error occurs and you might act to reduce or eliminate that error, such as, perhaps, by trying to get the cashier's attention, criticizing the person, or by leaving the store without buying anything.

However, occasions of abnormal behavior are fairly rare because your behavior and that of others is so routine in such settings. In fact, what you and others do is so routine and expected that the term *role* has been applied to characterize your behavior in certain settings. For example, as mentioned before, at a store, you are expected to perform the role of a customer if you want to buy something, and the person at the cash register is expected to perform the role of a cashier and exhibit the routine, expected behavior of that position. The same holds for your and other people's roles as well, as a parent, driver of a car, worshiper at a religious service, diner at a restaurant, worker, student, priest, pastor, rabbi, policeman, fireman, doctor, nurse, teacher, mailman, chairman of a meeting, basketball or football player, card player, bus driver, pharmacist, salesperson, professor, or holder of another social position. All such roles entail fairly routine, predictable, and expected behavior. Each role has its own reference script or program to be performed—similar to that of an actor in a play—although you and others can behave somewhat flexibly within the normal boundaries of the role being held and the behavior setting being occupied as long as perceptions of the reference program are not seriously violated.

Highlight 5.6 More Terms Reflecting Behavior Regularity

Behavior setting: "the geographical, physical, and social situation as it affects relationships and behavior" (VandenBos, *APA Dictionary,* 111); a standing pattern of behavior and the context or milieu surrounding the behavior pattern (Barker and Wright, *Midwest and Its Children,* 9, 45-46).

Standing pattern of behavior: "persisting, extra-individual behavior phenomena" (Barker and Wright, *Midwest and Its Children,* 7); "a bounded pattern in the behavior of persons, en masse; a discrete, extra-individual behavior entity, such as the game-playing of basketball team members.... Each standing pattern of behavior...has a precise and delimited position or location in time and space...such as a basketball game, a worship service, or a piano lesson" (Schoggen, *Behavior Settings,* 385).

Script: "a mental road map—containing the basic actions...that comprise a complex action. For example, the script for cooking pasta might be: Open pan cupboard, choose pan, fill pan with water, put pan on stove, get out pasta, weigh correct amount of pasta, add pasta to boiling water, decide when cooked, remove from heat, strain, place in bowl. Also called *script schema.*" (VandenBos, *APA Dictionary,* 820)

Social position: "In general, a social position is an identity that designates a commonly recognized set of persons.... each behave in characteristic ways. Physicians write prescriptions, school teachers lecture in classrooms, janitors sweep up, and so forth. Thus, each social position is said to exhibit a characteristic role." (Biddle, *Role Theory,* 5)

Job: "a specific duty, role, or function...a regular remunerative position" (*Webster's*, 650).

Rule: "a guideline or standard that is used to guide responses or behavior or that communicates situational norms" (VandenBos, *APA Dictionary*, 807). As in rules of a game and traffic rules.

When is Your Behavior More Unique?

Your behavior is not always regular and routine. Sometimes you do something quite different. When? There are three major times:

1. When what you are doing isn't producing the results you want,
2. When you face novel situations, and
3. When you are consciously learning to do new things.

Let's look at each of these situations in more detail.

When What You Are Doing Isn't Producing Desired Results

Sometimes things you are doing aren't working out as you would like. In such cases, you are not achieving the results that you desire. Your goals or other references are not being met.

For example, suppose you misplace your keys and look for them in the usual places with no luck. What happens then? If you really need those keys, you may become upset and behave somewhat randomly and perhaps even franticly—looking here and there, looking under things, looking in pockets and drawers, and so on. In short, your behavior becomes less routine.

Similarly, you may be having trouble with a friend, co-worker, or job assignment. You try doing something different to improve what is happening. Alternatively, you might even avoid the person or assignment that is causing the error. In other words, you do something less routine, atypical, and perhaps even original.

Why do you do such things? Looking at your troubles with Perceptual Control Theory in mind, we say that you do things to achieve your desired perceptions. You do different things until you either (a) hit upon something that works to reduce your error signals, or (b) given the situation you are facing, you change your reference levels so that they are more realistic (which is another way to reduce the error signals you experience), or (c) you simply give up trying. In any case, when things are not working out as you want, you do something different unless another more important goal is driving your behavior, such as happens when an abused woman doesn't leave her man because she loves him too much.

When You Face Novel Situations

Similarly, your behavior is often not routine when you face a new situation. In such situations, your regular behavior may not work to achieve the results that you desire, so you need to do something different. If you don't know what to do and other people are around, you may imitate what they are doing. Or you may ask for advice. Otherwise, you may try various other behaviors. For example, if you need to communicate with someone who does not speak your language, you may use gestures, draw pictures, or find someone to translate for you—these all being things that you might not normally do.

When You Are Learning To Do New Things

Likewise, when you learn to do something new, your behavior is less routine. At such times, you also consciously focus your attention on doing the new behavior involved. Over time, with practice, you usually perform what you have learned less consciously and more automatically.

For example, if you have learned to drive, at first you consciously attended to doing every new thing—from starting the car, putting it in gear, accelerating, steering to stay in the middle of the lane, maintaining an appropriate speed, braking, and parking. Later, with practice and experience, these behaviors became more routine, and you started to do them less consciously and more automatically.

In short, although most of what you do is relatively routine, occasionally your behavior is more varied. You sometimes do something different, even creative, such as when you prepare a new meal or write a book or article for publication (if you do such things). Similarly, you may do something else you haven't done before, such as driving to a new vacation spot or giving a novel speech—these also being different, non-routine behaviors. But even then, although your overall action may be new, the component acts and movements usually consist of fairly common behavior you have learned to do before, such as cooking on the stove, typing or writing words, driving a car, or giving a speech while standing, gesturing, and (if you are a good speaker) looking at your audience.[11]

Cultural Similarities and Differences

You also behave similarly in many ways to people around you. Such similarities are reflected in the concept of culture—groups of people tend to behave in similar ways. These behaviors may be like those of other groups or different. When the differences from other groups are noticed, they are sometimes described as cultural differences. As Highlight 5.1 indicates, the term "culture" refers to patterns of behavior and associated objects generally shared by members of a group. If you belong to a particular culture or sub-culture, you behave similarly, at least in some ways, to other group members.

For example, if you are an Anglo-American, you and others like you are somewhat similar. You all speak English or at least the American version of that language. You know the "Pledge of Allegiance" and stories such as "Cinderella" and "Little Red Riding Hood." You probably dress like most Anglo-Americans: for example, if you are female, you don't usually wear saris; and if you are male, you don't wear a turban. You probably value punctuality. You know how to function in common American behavior settings and behave properly in roles that you perform. If you are employed, your work-related roles are probably kept separate from your nonwork-related roles.[12] You share enough common ground with other Anglo-Americans (knowledge, beliefs, and expectations) that you can work with and carry out joint actions with them successfully.[13] Your individual habits and scripts for action are similar to those of many other Anglo-Americans.

In partial contrast, if you are a Hispanic living in the United States, you share a somewhat different set of beliefs and values along with other Hispanics. These probably include "an emphasis on the family, collectivism, a willingness to conform with the demands of people in authority, an acceptance of clearly defined gender roles (i.e. machismo and marianismo), a flexible attitude toward time, and a motivation to avoid conflict in interpersonal relationships (simpatia)," according to Eugene F. Stone-Romero and others at the University of Central Florida.[14] For example, compared to Anglo-Americans, you may be more willing to make financial sacrifices to attend family celebrations such as birthday parties, weddings, and baptisms for extended family members such as nieces and nephews. In contrast, the Anglo-American culture stresses that work is as important in a person's life as family.[15]

Other kinds of similarities in behavior are found in other cultures and subcultures. For example, in Chinese society, young people are expected to exhibit filial piety (*xiao*) including being respectful to elders.[16] If you are a Japanese woman, you are more likely than your husband to tutor and prepare your children for entrance examinations from kindergarten to university, lock up at night, take out trash, help others relax, select schools, pay bills, control the family's money, and make decisions about family savings.[17]

In other words, you and other members of the cultural groups and subgroups to which you belong routinely exhibit many similar behaviors. Why? In part, because you share many similar references such as expectations of role behavior, basic values, and norms. The same holds for members of other cultural groups. They also share many similar references, although some differ from your own.

How Similar are You to Others?

Although similar in many ways, you are also different and unique compared to other people. You have a different name, you differ genetically from others, your life experiences have been different, and you look somewhat different from others.

Although genetically unique, as is partially shown by your facial features, body structure, and hair and eye color, you are biologically similar to others in a number of ways.[18] For example, you have to eat, breathe, and maintain your body temperature within certain levels. That is, you, I, and everyone else are all autopoietic beings who have similar biological structures and functions. Our control systems work in similar ways and have a number of similar built-in references. These similarities result in similar behavior at times. As my mother used to say, "Even the Queen has to go to the bathroom."[19]

And, remember that if you belong to a cultural group or subgroup, some of your behavior is similar to that of others in your cultural group. If it wasn't similar, you wouldn't belong to that group. So, even though you are somewhat unique, your behavior is similar in some ways to other people.

But, even though you do many things similarly to other people, you do things differently as well. In fact, thousands of words exist to describe differences that are perceived to exist between people. Here are a few: trustworthy, dishonest; loyal, a traitor; helpful, obstructing; friendly, belligerent; courteous, rude; kind, spiteful; obedient, noncompliant; cheerful, gloomy; thrifty, wasteful; brave, cowardly; clean, dirty; reverent, and nonbeliever; as well as sincere, devout, single-minded, intelligent, dumb, frank, childlike, mature, energetic, and schizophrenic.

In short, you are similar to, as well as different from, other people—including queens and kings!

The Bottom Line

How Regular and Routine is Your Behavior?

To summarize, during a given day, you do things in a fairly limited number of behavior settings—for example, at your home, apartment, or dormitory; on streets and sidewalks; at your school or workplace; and perhaps in a store or two. In those places, you behave fairly routinely and normally. Most of what you do is done unconsciously and automatically, in familiar and regular ways. Sometimes, though, you may consciously observe what you are doing, think about it, and do something different—especially when what you are doing is not producing the results you want, when you face a novel situation, and when you are learning to do something new. However, such occasions, especially when you do something very different, are fairly rare. In fact, much of your behavior and that of other people is often so regular and automatic that it can be called "habitual." See Highlight 5.7 for more about habits.

Highlight 5.7 Habits

Habit: "a folk psychology term which may describe any of several levels of PCT hierarchical behavior: event, sequence, and program. In other words, habit describes a pattern of behaviors that seem to be the same or similar from all outward appearances."

(Klinedinst, "My Reply")

"Habits are learned sequences of acts that have become automatic responses to specific cues, and are functional in obtaining certain goals or end states."

(Verplanken and Aarts, "Habit, Attitude," 104)

"The essence of habit is an acquired predisposition to *ways* or modes of response. . . ."
"Man is a creature of habit, not of reason nor yet of instinct."

(Dewey, *Human Nature and Conduct*, 42, 125)

"Habits might be triggered by prior responses in a chain of responses; by environmental cues, such as time of day or location; by internal states, such as particular moods; and by the presence of typical interaction partners. . . ."

(Verplanken and Wood, "Interventions," 91)

"The basic problem . . . is that our consensus definition of habit defines habit as a *kind* of behavior (automatic, unconscious) but our theories employ habit as a *cause* of behavior. . . . You have simply provided a name for that kind of automatic behavior.

> Naming...is not the same as explaining..... Our theories of habit...need to stop viewing habits as causes and start looking elsewhere for the causes of automatic behavior—in the situation, in the body, and in the person's behavioral history...."
> (Maddux, "Habit, Health, and Happiness," 335-336)

Why Do You Behave This Way?

Most of the time what you do is what you do most of the time, as Townsend and Bever indicate in Highlight 5.3. Why do you act this way? Why do you act so regularly? It is because your routine, regular behaviors are effective in achieving perceptions that you desire and may even require in order to live and be happy!

Over time, you've learned to do things that work for you. You've learned to do things that achieve your desired perceptions and related references such as your goals. You've learned to do things that reduce errors of importance to you. And doing so in fairly routine, habitual ways has some big benefits. It saves a lot of time and effort thinking about what to do—a saving that makes your life easier and less complicated, helps you to perform tasks more quickly, permits you to give your attention to other matters, and helps you deal with many different controlled processes at one time—so many, in fact, that you are physically not capable of thinking about doing them all at once! Perhaps you have heard the old joke about someone being so dumb that they are not able to chew gum and walk at the same time. That is a gross simplification. Nearly everyone can walk, chew gum, and do many other things at the same time too. You can too, unless, perhaps, your legs are disabled. Thanks to your control systems and their automatic and routine operation, you can also breathe, remain standing or sitting upright without falling over, engage in conversation, and look at things—all at the same time without thinking about what to do![20] As William James wrote more than a century ago, "Habit simplifies the movements required to achieve a given result, makes them more accurate and diminishes fatigue."[21]

Besides making life easier for you, achieving your goals and other references automatically and regularly has great survival value. Imagine if you had to think about what to do each time you took a step, touched a hot object, avoided bumping into someone, or avoided hitting another car when driving. Instead, you are a survivor! And one reason is that your control systems enable you to do many things automatically, quickly, effortlessly, and at the same time.

Preview of the Next Chapter

Now that you have an idea how of regular your behavior is, when it varies, and why it does so, we will consider in more detail how your environment influences your behavior. The next chapter focuses on your immediate environment and how what is around you influences what you perceive and do.

Further Reading

For discussions of the pervasiveness of automatic and unconscious behavior, see:

- John A. Bargh and Tanya L. Chartrand, "The Unbearable Automaticity of Being," *American Psychologist* 54, no. 7 (July 1999): 462-479.
- Ran R. Hassin, James S. Uleman, and John A. Bargh, *The New Unconscious* (Oxford: Oxford University Press, 2005). A sampling of research and theory on the unconscious.

For an introduction to behavior settings and the patterns of behavior found therein:

- Roger G. Barker and Herbert F. Wright, *Midwest and Its Children: The Psychological Ecology of an American Town* (Hamden, CT: Archon Books, 1971). First published 1955; contains the original description of behavior settings.

Scholarly sources of information about social norms include:

- Michael Hechter and Karl-Dieter Opp, eds., *Social Norms* (New York: Russell Sage Foundation, 2001).
- Jon Elster, *The Cement of Society: A Study of Social Order* (Cambridge: Cambridge University Press, 1989). See Chapter 3: "Social Norms."

For more information about roles, see:

- Bruce Biddle, *Role Theory: Expectations, Identities, and Behaviors* (New York: Academic Press, 1979). An in-depth discussion of roles and role theory.

A good place to start reading to learn more about culture and behavior is:

- Harry C. Triandis, *Culture and Social Behavior* (New York: McGraw-Hill, 1994).

Endnotes

1. For example, see Bargh, "Automaticity of Everyday Life," 2, and "Reply to the Commentaries," 244.

2. PCT explains well how and why such variations in behavior occur, whereas behaviorist explanations are lacking in this regard. For details see Powers, *Behavior: The Control of Perception*, 2nd ed., 5-8 and Bourbon, "On the Accuracy," 45-46.

3. If we focus on your goals and other references, we can also say that the results or consequences of your behavior are even more regular than your behavior. For example, if you drive to your workplace, the result of driving there is that you actually arrive there every normal workday, even though the driving you do may vary from day to day because of traffic and weather conditions. Similarly, if your car is not working, you may go to work by bus, subway, taxi, or another means, but the consequence, arriving at work, is the same.

4. *APA Dictionary of Psychology*, 780.

5. Cummings, "Squat Toilets." An interesting study in this regard was conducted by David Nalbone and associates, who found that in a riverboat casino men's room, 90 percent of men washed their hands after urinating when someone else was seen in the restroom, but when no one else was apparently there, only 44 percent washed their hands. For details, see Nalbone and others, "The Effect of Social Norms."

6. Wood, Quinn, and Kashy, "Habits in Everyday Life," 1288.

7. Fiese et al., "A Review of 50 Years," 383.

8. Barker and Wright, *Midwest and Its Children*, 222.

9. Barker, *Ecological Psychology*, 164.

10. In such settings, participants perceive a difference between what is being observed and their reference for that perception that is important enough to take action to change the situation perceived, if possible, until it conforms to their reference.

11. Kirsch and Lynn, "Automaticity in Clinical Psychology," 508.

12. Stone-Romero, Stone, and Salas, "The Influence of Culture," 332.

13. Clark, *Using Language*.

14. Stone-Romero, Stone, and Salas, "Influence of Culture," 340.

15. Ibid., 341.

16. Yue and Ng, "Filial Obligations."

17. Triandis, *Culture and Social Behavior*, 50.

18. Even if you are a genetically identical twin, you and your twin have had somewhat different experiences, starting from the time when you were in your mother's womb,

where you occupied different positions and experienced somewhat different sensations. For example, one of you was closer to your mother's heart at different times and experienced its sound somewhat differently. You also have different names. Even if you are not a twin, your school, work, and play experiences have all differed to some extent from those of others, resulting in differences in your learning and behavior.

19. "The bathroom" in American English refers to "the toilet."

20. Doing things automatically also has disadvantages. One is that when you want to change your habitual behavior, your tendency to behave automatically may defeat your desire to change. You may find it hard to change your behavior. But more about this in Chapter 11, when we deal with how to change the things you do.

21. James, *Principles of Psychology,* vol. 1, 112.

PART III

Your External Environment

The following three chapters focus on your immediate and more distant environment, how you perceive that environment, and influences that your environment has on what you do.

CHAPTER 6

YOU AND YOUR IMMEDIATE ENVIRONMENT

The Big Picture

You are a product of your external environment that is partly a product of you. You exist because of your environment, and nearly everything you do is influenced by it and has been ever since you were conceived. In turn, you affect your environment and have affected it over your life, but to a lesser extent, since only parts of it are influenced by you.

For example, your parents conceived you. You developed in your mother's womb in ways affected by her health, experiences, and what she ate and did. Your brain developed networks of neurons based on your experience with the world—networks of neurons that affect what you perceive, how you think, and what you do. Your mother tongue was learned from those around you. Your family and others with whom you associated have affected your values, beliefs, and behavior. Schools and mass media have affected what you know about the world. If you work for a salary, your workplace has affected what you do when you work. You nearly always act in socially acceptable, normal ways when shopping, talking with others, driving, playing games, and doing other things with people. If not, your actions are corrected and influenced by those around you.

In turn, you use your environment to survive and do what you want to do. You are dependent on what is outside yourself for food, drink, and shelter. You use what is around you, such as clothing, fans, shade from trees, and central heating or cooling if you have it, to maintain your body temperature. You travel from place to place by means of public transport, cars, and perhaps bicycles, and do other things ranging from obtaining food (by using money) to eliminating your body wastes (by using toilets). You also interact with the people around you to do other things you want to do.

At the same time, you affect your environment. You affect the people around you by what you do and say. You affect the physical things around you by touching them, moving them, buying them, occupying them, eating some of them, and in other ways. In short, you interact with your environment and your environment interacts with you. Both you and your environment influence one another and what you do.

Highlight 6.1	Some Environment Variables That Affect Behavior	
• Advertisements	• Laws	• Policies
• Behavior settings	• Machines	• Pollution
• Building design	• Media	• Resources available
• Climate	• Money	• School size
• Consequences	• Music	• Signs and symbols
• Crowding	• Natural disasters	• Smells and odors
• Cues for action	• Noise	• Temperature
• Death of a loved one	• One's job	• Terrorist acts
• Drugs	• Opportunities	• Tools available
• Economic conditions	• Other people	• War
• Holidays	• Physical layout	• Weather
• Information	• Poison ivy	

What is Your Environment?

The exact nature of one's environment depends on whom you are focusing upon, since everyone lives in a different environment. Although some features of environments are similar, such as gravity, other aspects differ, even in the case of identical twins, such as fetal position in the womb, conversations that one has with others, and even one's name. So each person's environment is different, to a greater or lesser extent.

With this in mind we can say that once a person is selected, his or her *external environment* consists of those objects and processes outside of that person's body that affect the person or are affected by the person during a certain period of time.[1] The time period could be an instant, during a certain event, or during a period of the person's life. For example, we could focus on your environment at this moment (an instant), when you were born (an event), or when you were an infant (a period of your life).

Analyzing an environment can be extremely complex, depending on what level of analysis and how much detail one focuses upon.[2] In fact, some scholars think that it is impossible to identify all of the factors involved.[3] So, typically, analysts simplify the task and, given their purposes, focus only on what seems to be most relevant. For example, if you are upset, what made you upset? Or if you bought a particular product, what factors affected why you bought it? Or, more formally, during a judicial trial, what situation influenced your presumed behavior?

As Highlight 6.2 indicates, a number of words are closely related to the idea of environment. These include terms such as context, situation, setting, background, surroundings, and milieu. Although each has a somewhat different meaning to different people, each term refers to the same basic idea of things and processes that influence people or are influenced by them. Some definitions of these terms refer only to the external environment of a person, and some include the internal environment. For our purposes unless otherwise indicated, when we use the term "environment" or one of its related terms, we will be referring to your external environment, the part of the world outside of your skin.[4]

Highlight 6.2 Environmental Definitions and Synonyms

Environment: "the aggregate of external agents or conditions—physical, biological, social, and cultural—that influence the functions of an organism" (VanderBos, *APA Dictionary*, 334).

Context: "Context is that which environs the object of our interest and helps by its relevance to explain it. The environing may be temporal, geographical, cultural, cognitive, emotional—of any sort at all." (Scharfstein, *Dilemma of Context*, 1); "generally, the conditions or circumstances in which a particular phenomenon occurs" (VanderBos, *APA Dictionary*, 224).

Situation: "In *physical* and *biological* terms, a situation can be rather strictly defined as that part of the total environment that is available for sensory perception for a certain amount of time. To the physical and biological properties of places (churches, kitchens, clubs, ballrooms, buses, classrooms, laboratories, etc.) are attached *sociocultural* factors—norms, rules, roles, etc.—that contribute to a complete definition of an actual situation ...." (Magnusson, "Wanted," 14); [a situation] "is determined jointly by environmental and psychological factors" (Kelley et al., *Atlas,* 71).

Milieu: "the environment in general" (VanderBos, *APA Dictionary,* 579).

Your Behavior is a Function of You and Your Environment

It has been known for some time that behavior is a function of a person and his or her environment. Interestingly, in individualistic societies such as the USA, more emphasis tends to be placed on the individual when thinking about what causes behavior, than on his or her external environment.[5] In contrast, people in societies having a group or collective orientation tend to give more recognition to the external environment as a major cause of behavior. For example, when writing about a killer, an American newspaper might describe the person as mentally unstable, having a bad temper, and being a martial arts enthusiast, whereas a Chinese newspaper might indicate the easy availability of guns in the USA and that the person had strained relationships in his life, had recently lost his job, and was following the example of a recent similar slaying elsewhere.[6] Similarly, when thinking about atrocities such as those at the Abu Ghraib Prison in Iraq, where prisoners were abused and tortured by U.S. prison guards, most Americans tend to blame the soldiers involved rather than the environmental system of which they were a part—a matter explained well by Philip Zimbardo in his book *The Lucifer Effect: Understanding How Good People Turn Evil.*

In any case, your behavior is affected by both you and your environment interactively, so much so that it is almost meaningless to try to determine how much influence each has. Actually, both influence one another. Although your internal structure and organization affects how you behave and react to things, the relationship between you and your environment isn't one-way. If environmental conditions disturb an important perceived variable that you are controlling, you will act to restore your perception to its reference condition if you can do so and if other important controlled variables are not disturbed too much.[7] In this way, both you and your environment affect one another. For example, you are continually monitoring and controlling perceptions related to your references, such as the temperature of your body. If the temperature around you changes and you feel cold, you may put on more clothing, turn up the heat, snuggle up next to a loved one, shiver, or otherwise move and use your muscles to keep warm. In such ways, you affect your environment and your environment affects you.

Over the years, it has become increasingly clear that our body's structures and organization are affected by our environments, as indicated earlier by Highlight 4.1. For example, how your neurons are connected and networked is largely a result of your experiences from the time you were an embryo in your mother's womb until now. As a result, what you know and can do is largely based on experience you have had with your environment. For instance, your parents affected what you experienced and did as a child. Then, those experiences and what you did affected your neural development and organization. In turn, your neural organization affected what you knew and did, which then affected your parents and what they did. Similarly, if you work, a great deal of what you know and do there was learned at your workplace and perhaps at previous ones, too. In fact, estimates are that most of the knowledge and skills that we use when we work were learned on-the-job.[8] And, of course when you work, you affect your environment. Otherwise you wouldn't receive a paycheck! In short, both you and your environment continually interact, affect, and shape one another.

John Dewey once wrote, "Human nature exists and operates in an environment. And it is not 'in' that environment as coins are in a box, but as a plant is in the sunlight and soil."[9] You are embedded in your environment and dependent upon it to live and thrive. You were when you were a baby and your parents or others took care of you. Now, as an adult, you are still dependent upon others for food, clothing, and companionship to continue living as you do. You also affect that environment. You move furniture around, make other people happy and sad, and adjust the thermostats on your heating and cooling systems. This relationship is shown symbolically in Highlight 6.3 by the Chinese symbol of yin and yang.

Highlight 6.3 Yin and Yang: You and Your Environment

This symbol indicates your intimate relationship with your external environment. You, the large white area, are not only molded to your environment in ways that permit you to survive, but you are also surrounded by your environment. The black spot inside you and your general shape can be thought of as indicating how your environment has affected and continues to affect you. The shape of the black around you and the white spot, in turn, indicate how you have affected your external environment and continue to do so.

Within different environments, contexts, and situations, you behave in certain ways. For example, you behave as a shopper in a supermarket because of your perceptions and references, the way the supermarket is constructed, the way others are behaving in their roles as store personnel and shoppers, and your knowledge that if you behave too differently, you will be corrected, disciplined, not served, and perhaps even arrested. Why do you behave this way? Why do you behave as you do? Because doing so helps result in perceptions you desire—thus permitting you to live as comfortably as you can, given your present autopoietic makeup and the environment around you. In other words, *your behavior is a function of both you and your environment.*

Highlight 6.4 The Environment and Behavior

"We can recognize that all conduct is *interaction* between elements of human nature and the environment, natural and social." "Honesty, chastity, malice, peevishness, courage, triviality, industry, irresponsibility.... are working adaptations of personal capacities with environing forces."

(Dewey, *Human Nature and Conduct*, 10, 16)

"Human behavior is a function of the situations under which it occurs." "Stated simply, people behave differently in different contexts. Environmental factors are integral components of living systems' functioning. Behavior always occurs in a specific context and both must fit together to form a functional unit.... All 'habits' are linked to some kind of environment...."

> "A person sets out to produce some consequence(s) through his or her behavior. The person does a variety of things in and to a variety of contextual conditions until the intended consequence occurs, or until his or her attention and behavior are shifted to some other effort.... [The behavior] has coherence (it 'makes sense') if viewed as being organized to produce some consequence(s) in a relevant environment and... if viewed as a dynamically functioning self-constructing, adaptive control system."
>
> (Donald H. Ford, *Self-Constructing Living Systems*, 94-95, 145)

Your Natural Environment

Let's now consider one part of your environment, the "natural" part, and some ways that it affects what you and others do. Some aspects of your environment are so natural and a part of your life that you hardly think about them. And yet they affect you by physically limiting what you do, providing resources you need to live, and presenting opportunities to do things. For example, the force of gravity affects how you walk, sit, and deal with objects. That is, you walk, sit, sleep, and place things like plates and drinking glasses on the upper side of flat horizontal surfaces rather than on walls and ceilings. Similarly, the amount of daylight and lack of light at night affects you in many ways, including your sleeping patterns and activity levels. It's hard to do much in the dark, besides things like talking, listening, making love, and sleeping—that is, unless you turn on the lights. Likewise, where you walk is affected by natural structures such as uneven ground, heavy growths of vegetation, tree trunks, rivers, and cliffs.

Highlight 6.5 Body Temperature

> "... the safe upper limit for internal body temperature is roughly 41°C (106°F)."
>
> "Humans regulate internal body temperature rather tightly and generally defend it vigorously against it falling below 36°C."
>
> (Piantadosi, *The Biology of Human Survival*, 73, 112)

When the weather is hot, you tend to be less physically active, do less work, and relax more. When it is cold, you may move to a warmer place, wear warmer clothes, or adjust the air temperature around you by turning up the heat if you can. If it is really cold, you may shiver, exercise, or do some physical work to increase your body temperature. Similarly, other aspects of the weather such as rain, wind, and snow affect what we do.[10] Why? As explained before, we do what we do so to achieve our desired perceptions and related references—such as keeping comfortably cool and warm. If we become too hot or too cold, we may die!

The amount of oxygen in the air also affects what you do. If you change your altitude above sea level, your body will respond. If you go significantly higher, your breathing rate will increase to compensate for the decreased oxygen available. After a while, the concentration and number of your red blood cells will increase as your body adjusts to your changed environment. The blood flow to your brain and heart will also increase to maintain your oxygen supply to those vital organs.[11] In turn, your level of activity will be affected.

Natural smells and odors also influence you. When possible, you avoid disgusting smells and substances such as rotting meat and feces. You are attracted to things that smell nice. The same holds for sounds. Certain levels and kinds of noise, such as thunder, disturb you whereas the sound of water and waves gently lapping, may sooth and relax you.

Going back to the beginning of your life as an embryo in your mother's womb, you were also affected by that "natural environment." As you may recall from Chapter 4, your brain may have been affected by things as diverse as your mother's nutrition, illnesses, alcohol consumption, smoking, and drugs taken. Some parts and processes of your body, such as your height and hearing, may also have been affected.

To summarize, your natural environment has influenced you and your behavior from the very beginning of your life. Everyone else has been affected by his or her natural environment too.

Your Man-made Physical Environment

The man-made physical environment around you includes houses, schools, sidewalks, roads, cars, signs, furniture, pots and pans, books, electrical appliances, advertisements, machines, music, money, and packaged food. Homes and other buildings provide you with shelter and places to live and work. When the weather is bad, we stay inside if we can. If not, we may put on raincoats or use umbrellas to protect us. To sleep, we usually snuggle up in beds or fall asleep on sofas or other comfortable places. In such ways you make use of the man-made environment and are affected by it. Me too.

In fact, many buildings and other man-made spaces are designed to affect what we do and even how we feel. For example, cafeteria and supermarket layouts promote the selection of certain foods by placing them in more prominent positions. At workplaces, natural lighting and indoor and outdoor plants and trees may be arranged to help reduce stress and fatigue. Urban parks may be arranged this way too. In jails and other correctional settings, private rooms with outside windows, nicely painted walls using non-institutional colors, carpeted floors, and furniture with fabric upholstery help reduce violent behavior and vandalism as well as increase feelings of safety by inmates and staff.[12] Even the arrangement of chairs, tables, and desks can affect what people do, as many educators and trainers know and use to affect interaction patterns during their classes.

In addition, the man-made environment is often designed to provide cues, signs, and symbols aimed at communicating actions to take. Signs indicate what to do, the locations of places and things, and inform us about upcoming events. Written directions, instructions, and color coding help us perform appropriately. Sidewalks and roads indicate where to walk and drive. Clothes and uniforms may indicate one's role, status, position, and power (such as that of a policeman, doctor, or successful businessman) and help others to behave accordingly.[13] You, in turn, often notice these cues and use them to guide your goal-directed behavior. Such guidance and triggering of behavior often occurs unconsciously and is based on previous learning.

We typically benefit from the use of such cues, signs, and symbols. However, for people who have problems with their weight, gambling, smoking, or excessive spending, such cues may stimulate undesirable feelings or thoughts—such as an intense urge to eat something, or to gamble, drink, smoke, or buy things, for example, after seeing a lottery sales sign, strolling past a familiar bar, seeing a dessert tray at a restaurant, smelling cigarette smoke, or noticing an advertisement for a "Lowest Prices of the Season Sale."[14]

Highlight 6.6 Music and Consumer Behavior

Music has been shown to affect behavior in many ways. In a liquor store, for example, playing French folk songs has been shown to result in sales of more French wine, and playing German folk songs has been shown to result in sales of more German wine. In a restaurant, diners ate more quickly when fast music was played, and more slowly when slow music was played.

(North, Hargreaves, and McKendrick, "In-Store Music," 132.

We also use our man-made environment to do things more easily with less effort. Tools and objects ranging from paper and pens to computers, cell phones, and other technologies help us a lot. Calculators and computers simplify mathematical tasks. Maps and GPS devices help us find our way. Public transport, cars, bicycles, roads, sidewalks, and trails help us to move more easily from one place to another. In these ways we often use and are influenced by our man-made environment.

Your Social Environment: Other People

Overview

The people around you, called your "social environment," are an extremely important part of your life. Without them, you wouldn't be alive. They affect your survival as well as what you learn, know, and do.

Highlight 6.7 — Social Life

"If, following control theory, we take life's constant activity to be maintaining internal standards against threats from outside, then social life looms large, because other people comprise a very large part of the environment of most modern humans. Sometimes other people obstruct our purposes; sometimes they give us help."

(Runkel, *Casting Nets and Testing Specimens,* 144)

Your language, manners, roles, the "normal" behavior you exhibit, and the rituals and customs you follow, have all been affected by the people around you. As indicated in Chapter 4, you learned those things from others: by imitating them, by their teaching, and as a result of their reactions to what you have done. Through processes such as socialization, much of your behavior has been molded and shaped by other people. Techniques used can sometimes be quite subtle, as Highlight 6.8 indicates.

Highlight 6.8 — Examples of Techniques Used by Compliance Professionals to Influence What You Do

Robert B. Cialdini in his book *Influence: Science and Practice* 5th ed. mentions six major principles that compliance professionals (such as salesmen, fund raisers, and politicians) use to have you fulfill their desires for purchases, donations, votes, or assent to their requests. These "Weapons of Influence" are:

- **Reciprocation:** using the social norm that requires someone to repay what another person has provided—which is why advertisers give free gifts, samples, and meals; and salesmen and negotiators give concessions that then stimulate return concessions from you!
- **Commitment and Consistency**: the tendency of people to act consistently—which is why agreeing to a small request often leads to doing more, often much more, at a later time.
- **Social Proof:** finding out what other people are thinking or doing and then imitating them—which is a reason why canned laughter is used on TV; tip jars and church collection baskets are sometimes "primed" with money; artificial waiting lines are sometimes created; and testimonials, "largest selling" and "fastest growing" ads exist.
- **Liking:** People prefer to say yes to people they know and like—which is a reason why you often see well-known, popular figures and good-looking people in advertisements, and why requesters may flatter you or indicate they are similar to you, since we tend to like such persons.
- **Authority:** obedience to proper authority—and so salesmen, actors and others often wear business suits or white coats (in dental and medical advertisements) to look more authoritative, use prestigious titles such as Dr. or Vice President, and pose as experts.
- **Scarcity:** More value is given to scarce items—which is why stores and other businesses have limited sales and exclusive engagements, limited numbers of items available, and one-time offers; and why activists flaunt imagined or sometimes real loss of freedom or reduction of other valued services to persuade you to act as they want.

Here are some examples (from Wikipedia, 2009):

- **The door-in-the face technique** involves starting with an extreme request that is sure to be turned down and then retreating to a smaller request that the requester had in mind all along. For instance: *"Will you donate $1000 to our organization?* [Response is no]. *Oh. Well could you donate $10?"*
- **The foot-in-the-door technique** is a compliance tactic that involves persuading a person to agree to a large request by first having the person agree to a more modest request. For instance: *"Can I go over to Suzy's house for an hour?"* followed by *"Can I stay the night?" "Would you sign this petition for our cause?"* followed by *"Would you donate to our cause?"*

Cialdini's book contains many more examples and much more information about such techniques of influence and persuasion. It is well worth reading.

Since you were born you have spent so much time with other people and are so dependent on others to survive and do the things you want to do, it is perhaps not surprising that those around you have had so much influence on you and what you do. What follows can help you better understand how such influence occurs.

Social Control

James J. Chris writes, "The study of social control is the study of how society patterns and regulates individual behavior."[15] He points out that there are many mechanisms and procedures in place for attempting to affect the compliance of individuals and groups to some ideal standard of conduct. Such ideal standards, of course, are what Powers calls references and for which many other terms have been used such as values, duties, goals, and preferred states, as Highlight 3.4 indicates.

You may recall from Chapter 4 that people around you attempt to influence your behavior to help ensure that what you do is acceptable to them. In other words, the people in your immediate environment are observers who perceive what you do. If your behavior differs from what they consider to be appropriate behavior, they may take action to change what you are doing until what they perceive is more in keeping with their references. They may look harshly at you, criticize you, correct you, ostracize you, or even send for the police to deal with you. Alternatively, if what you are doing matches their references for appropriate behavior, they may smile at you, compliment or praise you, give you a reward, or otherwise act in a way that is pleasing to you. Whether they take action or not depends on what they perceive, the strength of their error signals, if any, and if they feel that doing something will have a desired effect.

When two or more people with similar purposes seek to control the same thing in the environment, the resulting collective control is usually more powerful than that of only one person, according to Kent McClelland, a sociologist.[16] In other words, people working together to achieve the same goal, are nearly always more effective than is only one person when attempting to influence other people, things, and events. That is why group action is often much more effective than individual action. Few people alone are willing to stand up to groups and face possible consequences of disapproval, ridicule, criticism, and rejection. Are you? That is a reason why other people have been so effective in influencing what you do. That is an important reason why you act so normally, at least within the bounds of reasonable social acceptance, when you are with groups of other people.[17] It also explains why when you are with different groups who value things differently, you may act as they do when you are with them.

McClelland also points out that rules, norms, laws, customs, mores, and codes of conduct can all be regarded as collective control processes. He writes, "Some activities—engaging in conversation, holding a meeting, playing competitive games, making love or war—are collective control processes by their very nature. Other collective control processes provide the floor of expectations on which we base our independent moves."[18] Some of his other thoughts about collective control are indicated by Highlight 6.9.

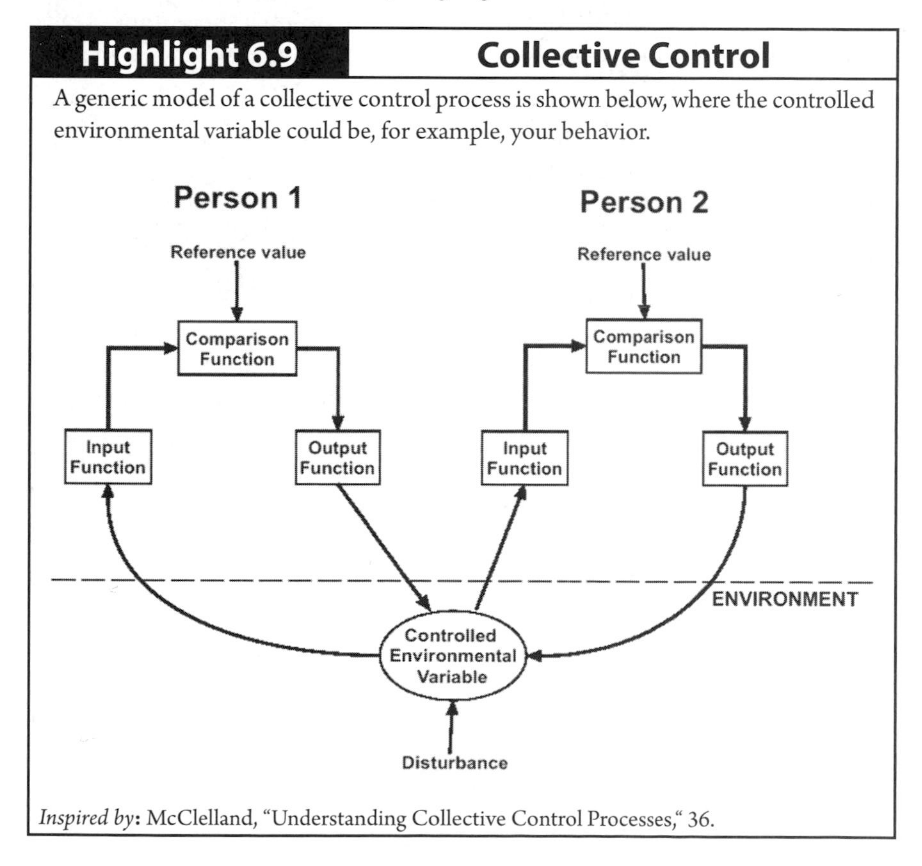

Highlight 6.9 Collective Control

A generic model of a collective control process is shown below, where the controlled environmental variable could be, for example, your behavior.

Inspired by: McClelland, "Understanding Collective Control Processes," 36.

As shown by Highlight 6.9, different people sometimes try to influence the same thing in their environment—for example, to move a refrigerator or to have you behave in a certain way. When the references of the different people are similar, their efforts are often more powerful than that of only one person. Therefore, they use their joint efforts to move the refrigerator or to try to have you do something.[19]

If your references and desired perceptions are similar to others and if what they want does not conflict with important references you have, then you can be easily influenced to do what they want. However, if your references are incompatible with theirs, conflict will occur. In either case, what you do depends on you as an autopoietic being and thus on the structure and organization of your control systems—unless those trying to control you use overwhelming threats or force! That is why Powers and others make the statements about trying to control other people as shown in Highlight 6.10. That is why you cannot really control other people. At best, you can influence what others do—preferably by working with them in ways that will help them achieve or maintain their desired perceptions. In other words, "our actions reflect what we want," as the family counselor Edward E. Ford stated so well.[20]

Highlight 6.10 **Is Control of Other People Possible?**

It appears that while people can influence one another, they cannot really control what someone else does—at least without producing conflict. This is because what someone does is controlled by their internal structure and organization, not by their outside environment—as explained in this book from Chapter 1 onwards. Here are some related thoughts:

"Control of [other people's] behavior….is simply inconsistent with the facts of human nature." "People cannot get inside each other's brains to operate the control systems there, and those control systems are what cause behavior."

(Powers, *Behavior: The Control of Perception*, 271)

"It is not so much a matter of what to do with a person … as it is a matter of getting 'what I want and what you want' to match."

(Robertson and Powers, *Introduction to Modern Psychology*, 159)

"People control their own experiences. The only way you can truly force them to behave as you wish is through the threat or actuality of overwhelmingly superior physical force…."

(Powers, *Making Sense of Behavior*, 122)

"You can't really change another person, only yourself."

(E. Ford, *Love Guaranteed*, 40)

To learn more about such thinking, you are referred to Powers, *Behavior: The Control of Perception*, 250-272; Powers, *Making Sense of Behavior*, 91-124; and Marken and Carey, *Controlling* People, 73, 82-5, 97, 143.

When the references of systems are incompatible and the action of one person prevents or attempts to stop another person from getting what he or she wants, then conflict occurs. Such is sometimes the case with a parent and a child—when the parent wants one thing and the child another. For example, the child might say, "I want to see the movie with my friends." The parent may then say, "No, you can't go. You need to do your homework!" Or, an adult child might say, "I want to marry him!" The parent may reply, "No, he is a disaster! You'll ruin your life and our family's reputation."

The same holds for groups. When one group of persons' references, such as goals or values, are incompatible with those of another group and they try to control the same environmental variable, conflict will occur between those groups. Politics is an area where such conflict can be frequently seen. Wars are another, more extreme, example.

How such conflicts play out depends on the internal dynamics of those involved and on the resources that each side may bring to bear on resolving the conflict and other disturbances that affect what is going on. To know more about social conflict and its resolution, besides the references shown in Highlight 6.10, a book by Marken and Carey titled *Controlling People* is worth reading.

Behavior Settings: A Combination of the Physical and Social

You may recall that behavior settings are places where your behavior is fairly routine. Such settings often consist of combinations of the different kinds of environments we have been discussing—natural, man-made, and social. Examples include parks, roads and highways, post offices, banks, supermarkets, workplaces, barber and beauty shops, school classes, church services, birthday parties, baseball games, and formal meetings. Every day, you usually move from one behavior setting to another, spending time in each, and behaving in general ways like most other people in those settings. You adjust your behavior to the setting you are in. So do other people.

Highlight 6.11 Behavior Settings and Behavior

"Our findings indicate that most of the environments in which people live are homeostatic systems that maintain their characteristic patterns (including the behavior of the inhabitants) within preset limits by means of control mechanisms."

"The behavior of persons who move between behavior settings conforms to the pattern prevailing in the setting they currently inhabit."

(Barker et al., *Habitats*, 42, 194)

"Behavior settings require conformity of their inhabitants, but they do not require uniformity."

"Some aspects of the behavior of different persons within the same behavior setting may differ widely: one person may enter a drugstore to buy medicine for a friend, another may enter to buy poison for an enemy; one person may go to church for spiritual satisfaction, another for social advantage.... Yet all these people will conform to the standing pattern characteristic of behavior in the setting. In other words, the content and structure of a person's own psychological world, his life space, are by no means determined by the behavior setting."

"Within a behavior setting there are routes to goals that are satisfying to the inhabitants. A setting exists only so long as it provides its inhabitants with traversable routes to the goals their own unique natures require."

(Barker, *Ecological Psychology*, 29, 167, 195)

When driving, you stay on the proper side of the road, stop at red traffic lights, and generally obey traffic regulations. If you don't, other drivers may blow their horns at you; you may have an accident; or a policeman may give you a warning, a ticket, or arrest you. Similarly, if you are at a religious service in most churches, you sit or stand facing the pastor or priest, are quiet when he or she talks, and otherwise act appropriately and somewhat predictably. If not, you may be corrected, shunned, or removed by those around you. The same holds for other public settings. Of course, in more private settings, such as your bedroom and bathroom, you have more freedom in how you behave. But even then, your behavior is fairly routine and constrained by the setting you are in. For example, you probably don't dine with friends in the bathroom, or take a bath in the dining room.

People in behavior settings consciously maintain those settings to match their desired perceptions. Doing so seems to be why Barker called these environments "homeostatic systems." Such maintenance is usually done by those who are in charge of the settings— the shopkeepers, parents, pastors, and other managers. Balanced stable states are maintained or restored by correcting or removing troublesome parts that produce important error signals. In a store, if the shelves are emptied, managers will ensure that they are restocked. If a shelf is broken, it will be repaired or thrown away. If a sales clerk makes mistakes, he or she will be corrected, trained, or fired.[21] You too will be dealt with if you behave too strangely or inappropriately. Bystanders and others in the setting may also take part in correcting deviations from their references. For example, in a store, a customer may report a thief who is stealing. Or, bystanders may tackle someone who is shooting other people.

In behavior settings where you perform a critical role, such as that of a parent, teacher, manager, or chairperson of a meeting, you will also act fairly routinely and predictably. If deviations occur within the setting, such as children misbehaving, equipment being broken, or people not following the rules, you may take action to correct the situation. In such ways, you both affect and are affected by the behavior setting you are in.

Why do you do these things? As you may realize by now, your behavior is explained well by Perceptual Control Theory. That is, when you are in a behavior setting, your perceptions are compared to your references. When important differences occur, you may take action to ensure that those differences are reduced or eliminated. As a shopper, you might experience perceptions that lead to error signals—such as empty shelves or unsatisfactory service—in which case you may complain or go to another store to meet your needs. Similarly, other shoppers behave according to the references they have. In such ways the stable nature of behavior settings is maintained by you and others who work in or use those settings.

Another aspect of behavior settings that affects what people do concerns their size and the opportunities they provide. For example, smaller schools with 100 to 200 students provide more opportunities for learners to become involved in activities such as plays and sports teams than do larger schools. In fact small schools almost demand that each student take part if many school activities are to take place. In contrast, a large school, such as a public high school with 1000 or more students, provides relatively fewer opportunities to take part in school plays, the football team, the basketball team, and other activities. As a result, more students in large schools are marginalized. The same holds for undermanned organizations such as small businesses or small chapters of professional associations. In such cases, opportunities for performing a variety of tasks are available for each employee or person involved. Actively participating in a variety of jobs or tasks by individuals may even be essential if a small organization is to survive. Such opportunities for action and limitations on what can be done are often unrecognized, even though they affect what you and others do.

Lastly, much of your memory is also linked to behavior settings. That is, such settings often remind you of proper ways to behave. Thus, your recall of norms, expectations, and roles is often linked to the setting and situation you are in. Much of this occurs automatically, without your even thinking about it. We'll discuss this more in a few pages when we come to "Priming Effects."

Other Interactions with Your Environment

Genes, Neurons, and Environmental Influences

Your genes and their functioning are affected by your environment and have been ever since you were conceived. Your environment affects the turning on and off of your genes and the proteins that they make, including hormones, neurotransmitters, and resulting connections between your neurons. This influence occurs either directly by chemicals, radiation, and gravity, or less directly by triggering your sensory neurons and related perceptions. Highlight 6.12 describes how these environmental effects occur. As indicated before by Highlights 4.1, 4.10, and 4.11, even fairly small environmental influences can have profound effects on your body's structure, function, and resulting behavior.

Environmental effects on neurons occur much more rapidly than they do on genes. As the research psychologist Gary Marcus indicates, genes work on a time scale of seconds or minutes whereas neurons act on a scale of hundredths of a second. Genes mainly lay down and adjust neural circuitry, but they are not involved much in the rapid running of the nervous system. "Genes build neural structures—not behavior," according to Marcus.[22] See Highlight 6.12 for some of his other thoughts about how genes operate.

In addition to stimulating your sensory neurons, your environment may also influence what you attend to. That is because your attention is usually automatically directed toward sudden changes that are sensed in your environment, as well as on moving objects and unexpected or novel events. Your attention is also engaged by especially large, bright, or loud objects and by external stimuli to which you associate special significance. For example, loud noises automatically attract your attention. So do moving objects such as something coming towards your face, other movement (like on a TV screen), events such as an unexpected taste or an unexpected noise at night, significant sounds such as when someone says your name, and other perceptions having significance such as the sight of blood, snakes, and spiders.

Highlight 6.12 Environmental Effects on Genes

"Although every cell contains a complete copy of the genome, most of the genes in any given cell are silent.... Each gene has an IF and a THEN, a precondition (IF) and an action (THEN).... Rather than being static entities that decide the fate of each cell in advance, genes—because of the regulatory sequence—are dynamic and can guide a cell in different ways at different times, depending on the balance of molecules in their environment."

"The essential point is that genes are IFs rather than MUSTs. So even a single environmental cue can radically reshape the course of development."

"Because each gene has an IF, every aspect of the brain's development is in principle linked to some aspect of the environment; chemicals such as alcohol that are ingested during pregnancy have such enormous effects because they fool the IFs that regulate genes that guide cells into dividing too much or too little, into moving too far or not far enough, and so forth."

(Marcus, "Making the Mind," 139-140)

Such focusing helps ensure that you notice important events, objects, and features of objects.[23] Businesses rely on this when they attempt to direct your attention to their products and services and hopefully persuade you to buy them. Advertisers and TV producers are well aware of techniques that grab your attention and use them frequently.

Highlight 6.13 Some More Definitions

Attention: "a state of awareness in which the senses are focused selectively on aspects of the environment and the central nervous system is in a state of readiness to respond to stimuli."

Priming: "in cognitive psychology, the effect in which recent experience of a stimulus facilitates or inhibits later processing of the same or a similar stimulus."

(VandenBos, *APA Dictionary*, 82, 731)

Priming Effects

An important and usually unrecognized influence of the environment on what you do is "priming." Let me explain.

Your sensory neurons are constantly stimulated by your environment. These neurons, in turn, stimulate other neurons and networks of neurons. Once stimulated, neurons tend to be more sensitive for a period of time and more easily stimulated again. Such sensitivity affects networks of neurons such as those associated with your perceptions, goals, concepts, ideas, stereotypes, and actions. This sensitivity helps you react more quickly to your environment.

The development of such neural sensitivity is called "priming." That is, *priming* refers to the recent activation of a neural network and the resulting facilitation of related perceptions, thinking, and action as well as the inhibition of unrelated perceptions, thinking, and action.

Here are some examples of priming effects:

- Slow music tends to result in people staying longer in stores, supermarkets, and restaurants, whereas fast music tends to increase turnover rates.[24]
- If you are consciously looking for a certain object, you are much more likely to notice it than when you are not primed to do so.[25]
- Exposing people to words related to "kindness" is likely to cause some to later view someone else as more kind.[26]
- A priming stimulus of "generous" can affect one's impressions of another person, trigger altruistic goals, and increase the likelihood of generous behavior such as when asked to donate to a charitable organization.[27]

In some cases, your actions may be inhibited by priming effects. Such inhibition is sometimes called *negative priming*.[28] Negative priming occurs, for example, when you work on a task and sensations not relevant to the task are either not noticed or take longer to notice. For example, you can consciously experience examples of this effect by searching online using the words "selective attention test" and then looking at some of the links.[29]

In short, your body and its control systems are constantly being stimulated and primed by your environment. This, it seems, is one reason why your brain, which accounts for only about 2 percent of your weight, is constantly active and accounts for about 20 percent of the total energy that you use.[30] Such stimulation triggers the rapid activation of numerous control systems within your body that permit you to easily maintain your balance, walk without stumbling, sit without falling over, drive to destinations without hitting other cars or going off the road, converse with others, eat by smoothly moving food to your mouth, act appropriately in different situations, and do most other things.[31]

Highlight 6.14 **Priming Effects**

"A large number of studies in the social cognition literature... have demonstrated that the activation or priming of a particular concept in memory increases the likelihood that this concept will influence subsequent judgments and behaviors."
(Petty, "Subtle Influences," 134)

"When observers see a motor event [behavior] that shares features with a similar motor event present in their motor repertoire, they are primed to repeat it."
(Rizzolatti and Craighero, "The Mirror-Neuron System," 180)

"The main moral of priming research is that our thoughts and our behavior are influenced, much more than we know or want, by the environment of the moment."
(Kahneman, *Thinking Fast and Slow*, 128)

Priming may even be a basis for copying other people's behavior. Art Dijksterhuis states, "The mere perception of an act elicits the tendency to engage in the same act."[32] For example, we often mimic the postures, gestures, facial expressions, speech, and emotions of others.[33] For instance, if you are talking with someone who has folded his arms in front of him, you may fold your arms, too. If someone near you yawns, you may yawn, also. When you hear laughter, you may laugh—which is why many TV sitcoms use "canned laughter" to help you feel that what they are showing is funny. Such priming and resulting mimicry, when it occurs, is another example of environmental influence on your behavior.

Another type of priming is sometimes called *contextual priming*, which refers to the activation caused by contexts upon neural networks associated with that context. This seems to account for some of the routine behavior you exhibit in behavior settings. For example, when you go into a restaurant for a meal, your reference and role as a patron is primed along with appropriate sequences of behavior in that context: for example, wait to be seated, read the menu, give order to waiter or waitress, eat, receive and pay the bill, and leave. The same holds for other contexts, such as classrooms, churches, and gasoline stations. However, you are primed differently when you walk through a supermarket than when you are at work, at home, or walking through the woods. At the supermarket rather than at work, you are more ready to look for and buy things that you need. At your workplace, you normally wouldn't look for the shortest line at a checkout counter. At home, rather than at work, you are more likely to take your clothes off, possibly put on pajamas or a nightgown, and then sleep. In the woods, you are more likely to detect motion than at a supermarket.[34]

Highlight 6.15 **Priming is Unconscious Adaptation to Your Current Environment**

"The unpredictability of the future, as it relentlessly approaches us, requires us to be continuously *reactive* to unfolding events. Because we can't know with any degree of certainty what will happen in advance (in most natural situations), we have to react and adapt to what is currently going on—and the evolved design of our minds causes the on-line presence of these events and objects and people to automatically activate our internal representations of them. With the activation of the representations comes, concomitantly, all of the internal information (affect, goals, behaviors, knowledge) relevant to responding back to the current environment...."

(Bargh, "Free Will Is Un-natural," 140)

Contextual priming has other subtle effects. For example, one study showed that 2 percent more people voted for increased funding on a school funding initiative when the polling place was at a school, compared to people who voted in non-school settings.[35] Another study showed that both children and adults consumed more snack foods after exposure to food advertising on TV, with children eating 45 percent more than others who had seen advertising that was not about food.[36]

Advertisers are aware of priming effects and use it to their advantage, such as the TV advertising effect on eating behavior mentioned above. Others include the placement of advertisements in magazines. For example, advertisements emphasizing a certain product feature such as air bags may benefit by having an adjacent article such as a crime story that can prime the product's quality of safety.[37] Even the time of year may serve as an effective contextual prime to influence sales. That is why during the summer, there are more ads for ice cream, suntan lotion, soft drinks, and swimwear; at the start of the school year, more ads appear for children's shoes and school supplies; and during winter more ads are seen for canned soup, cough medicine and cold preparations.[38]

Finally, you should realize that you can also prime yourself in helpful ways. One technique is to make a plan and form an "implementation intention" to do something to help achieve your plan's goal when you encounter a specific situation or when you notice a specific cue. That is, make an *if-then* plan: "*If* Situation X is encountered, *then* I will do Y." The situation or cue, when noticed, will then trigger the goal-related action on your part if all goes according to plan.[39] For example, you may decide if someone says certain things, you will respond in a particular way. Or if you are watching your weight, you may decide that if you go to a buffet or someone's house to eat, you will take only one modest helping of food and will not have a second helping. We'll discuss implementation intentions more in Chapter 11 when considering how to change your behavior. But for now you should realize that your intentions can prime your behavior for automatic activation by environmental cues.[40]

Highlight 6.16 Priming and Aggression

"A considerable number of studies examined the potential effects of violent media exposure on aggression. Overall, this research, whether conducted on violent television, films, music, or video games, consistently found that exposure to violent media is associated with aggressive behavior...." "...media violence can produce short-term increases in aggression by increasing the accessibility of aggressive thoughts, making a person feel angry or hostile, or by increasing the person's psychological arousal."

(Swing and Anderson, "Media Violence," 103)

"Briefly, priming is the process through which activation in the brain's neural network spreading from the locus representing an external observed stimulus excites another brain node representing aggressive cognitions or behaviors.... These excited nodes then are more likely to influence behavior. The external stimulus can be inherently aggressive, e.g., the sight of a gun..., or something neutral like a radio that has simply been nearby when a violent act was observed.... A provocation that follows a *priming* stimulus is more likely to stimulate aggression as a result of the priming. While this effect is short lived, the primed script, schema, or belief may have been acquired long ago and in a completely different context."

(Huesmann, "Imitation," 258)

"When one thought is activated, other thoughts that are strongly connected are also activated. Immediately after a violent film, the viewer is primed to respond aggressively because a network of memories involving aggression is retrieved."

(Tedescchi and Felson, *Violence*, 114)

To summarize this section: Your environment as well as your thoughts and feelings are constantly priming your neural networks and preparing you for action. That priming doesn't cause you to act in certain ways. But by sensitizing your neural networks, you are more likely to perceive and do certain things, and more quickly too. The same holds for other people, as Highlight 6.16 indicates.

Learning

Learning is nearly always based on experience you have with your environment. For example, all the ways of learning discussed in Chapter 4 are based on contact with your environment. Three important details follow:

Environmental Regularities, Relationships, and Patterns

You notice and learn from regularities, relationships, and patterns that you perceive in your environment. That is, you learn to associate objects and events that occur together at the same time or that follow one another. Perceived associations of objects and events occurring in your environment are the basis of much that you have learned. Such regular relationships are also a basis for your habits.

Habits

The situation specific nature of your habits points to another way that your immediate environment affects what you do. In order to achieve your references in specific situations, you tend to do things in the same general ways. In such situations, you often exhibit the similar patterns of behavior that are called "habits" by many people.[41]

You developed habits because in those situations and contexts the ways you behaved achieved your goals and other references. By repeatedly doing similar things in those situations, you learned that those actions were effective in helping you achieve what you wanted. They helped you obtain or maintain desired perceptions concerning a goal, homeostatic condition, or other reference. And so, your habitual behaviors are intimately linked to your environment, since they:

- were learned as a result of interactions you had with your environment,
- are usually initiated by certain situations and related cues that you encounter, and
- help you to achieve your goals and other references in those kinds of situations.

Highlight 6.17 Habits and Your Environment

"All 'habits' are linked to some kind of environment...."

(Ford, *Humans as Self-Constructing Living Systems*, 95)

"Habitual responses are likely to occur with minimal thought and effort to the extent that the contextual features integral to performing the response and one's behavioral goals are similar across time and setting."

(Ouellette and Wood, "Habit and Intention," 55)

"Psychologists have been fairly unanimous in adopting a... conceptualization of habit as learned sequences of acts that become automatic responses to specific situations, which may be functional in obtaining certain goals or end states.... Habits thus comprise a goal-directed type of automaticity, which may be consciously instigated...."

(Verplanken, Aarts, and Knippenberg, "Habit, Information Acquisition," 540)

Let's consider some examples of habits—also called "personal mannerisms," "styles of behavior," "default ways of behaving," and "scripts" by some people.[42] Habits range from behaviors and actions such as brushing your teeth in the morning and other activities that you tend to do such as listen to the radio or look at particular TV programs, to more complicated behaviors such as the way that you habitually drive to work (if you work and drive) and what you routinely do in behavior settings. As stated by the philosopher John Dewey and indicated in Highlight 6.4, "All virtues and vices are habits"—habits that "are working adaptations of personal capacities with environing forces."[43] In other words, habits are linked to the existence of certain environmental conditions. They can also be changed by modifying aspects of the environment that support those habits[44]—this being something that we will discuss more fully in Chapter 11 which deals with how to change your behavior.

Critical Periods of Learning and Development

During certain periods of your life, you also learned some things quickly as you interacted with your environment. When those periods passed, those kinds of learning or development did not occur as well or as easily as before and still do not. Language learning is an example. When you were young, you learned one or more languages very quickly without much effort or thinking—about two to six words a day between the ages of two to six years.[45] During and after adolescence, your ability to learn languages became less efficient, with new languages, if any, being learned more slowly, with more effort, and probably with an accent.

Other examples related to "critical" or "sensitive periods," as they are called, are not quite as obvious. However, as indicated by Highlight 4.1, a lack of environmental stimulation during infancy can have lasting effects upon a person's neural development and intelligence. Other aspects of one's immediate environment can have effects, too, especially during embryo development and early infancy. These include exposure to certain diseases, maternal smoking and drinking, nutritious foods, and toxins including recreational drugs. Effects of such environmental experiences and agents include everything from perception and mental retardation to one's height as Highlights 4.1 and 6.18 indicate.

Highlight 6.18 — Sensitive Critical Periods

"Many human capabilities depend critically on experience gained during early life. These capabilities range from fundamental capacities, such as stereoscopic vision, visual acuity, and binocular coordination, to high-level capacities, such as social behavior, language, and the ability to perceive forms and faces. In each case, normal experience during a restricted period in early life is essential for the normal development of the capacity."

(Knudsen, "Early Experience," 521)

In short, exposure to environmental events and conditions during critical periods, such as a caring parent or guardian, visual experiences and sounds, and certain substances are important for some types of learning and development to occur. A lack of such events and conditions can stunt or divert learning and development in less desirable ways.

Body Structures and Functions

As you now know, your body's structure is partly a result of your environment. So is the way you function. Here are a few more examples of such relations:

- A lack of iodine intake is the commonest cause of mental deficiency worldwide.[46]
- Nutritional supplements have been shown to reduce levels of violence in prisons.[47]
- Drugs have both beneficial and harmful effects—ranging from preventing and curing disease to affecting judgment, reaction times, and coordination (alcohol), causing hallucinations and paranoia (amphetamines),

creating addictive cravings (cocaine, crack, heroin), feeling giddy and confused (inhalants), impulse control problems (aripiprazole), and intensifying and distorting senses (LSD).[48]

Highlight 6.19 **Lack of Iodine: An Example of Environmental Impact**

"On a worldwide basis, iodine deficiency is the single most important preventable cause of brain damage."

"When iodine intake falls below recommended levels, the thyroid may no longer be able to synthesize sufficient amounts of thyroid hormone. The resulting low level of thyroid hormones in the blood... is the principle factor responsible for damage to the developing brain and other harmful effects.... The most critical period is from the second semester of pregnancy to the third year after birth...."

"As a result, the mental ability of ostensibly normal children and adults living in areas of iodine deficiency is reduced compared to what it would be otherwise.

Thus, the potential of a whole community is reduced by iodine deficiency. Where the deficiency is severe, there is little chance of achievement and underdevelopment is perpetuated. Indeed, in an iodine-deficient population, everybody may seem to be slow and rather sleepy. The quality of life is poor, ambition is blunted, and the community becomes trapped in a self-perpetuating cycle."

(World Health Organization, *Assessment of Iodine Deficiency,"* 1, 6-7)

Environmental Choices and Shaping

You also choose and shape some of your environments. In turn, you are shaped, chosen, and rejected by some of those environments. From infancy onwards, you have influenced your environment, in part, by crying as an infant, by talking with people, by buying things, by doing things such as rearranging furniture, and perhaps by gardening, voting, and helping, harming, or otherwise affecting other people. As you became older, you increasingly chose and changed environments of which you were a part. For example, even now if you have the opportunity, you may choose where to work; where to vacation; which clubs, associations, and religion to belong to; and who your friends are. You may also leave certain environments, especially when they are not satisfying your goals and you may be able to better achieve those goals elsewhere.[49]

Some parts of your environments have also chosen you. These may include your family if you were adopted, your spouse if you are married, your employer if you are an employee, the police if you were ever ticketed or arrested, as well as peer groups, clubs, sports teams, and college.

Entrance requirements are one way that colleges, social clubs, and professional associations choose their members. Many formal groups also have constitutions or bylaws that describe how a member can be removed. In such ways, your environments have affected you in terms of whom you associate with, what you do, do not do, know, and believe. And, in turn, you have also sometimes chosen and affected your environments.

Highlight 6.20 Environmental Selection

"People actively select and create their own environments. Environments also actively select and 'create' their own people."

(Moos, *The Human Context*, 398)

Your Core Affect

Lastly, your "core affect" is influenced by your environment. Core affect refers to your overall feeling at a particular moment such as being calm, sad, gloomy, upset, tense, excited, happy, serene, and contented.

External influences on your core affect include your physical environment, such as the weather, odors, and noise; and social factors, such as who is nearby and the type of situation you are in.[50] And so, how you feel is partly affected by your environment. And how you feel affects what you do.

Preview of the Next Chapter

After reading this chapter, you should have a better understanding of how you and your environment influence one another. The next chapter focuses on what your environment is really like, how you perceive it, and why your perceptions are so important and sometimes different from those of other people.

Further Reading

For information about techniques of influence and persuasion, an easily readable and informative book is:

- Robert B. Cialdini, *Influence: Science and Practice*, 5th ed. (Boston: Pearson, 2009).

An important, entertaining book that discusses environmental effects on children, especially those of peers versus parents, is:

- Judith Rich Harris, *The Nurture Assumption: Why Children Turn Out the Way They Do* (New York: Touchstone, 1998).

For more information about behavior settings, see the classic book:

- Roger G. Barker and Herbert F. Wright, *Midwest and Its Children: The Psychological Ecology of an American Town* (1955; repr. Hamden, CT: Archon Books, 1971).

For a scholarly introduction to control system theories in sociology, see:

- Kent A. McClelland and Thomas J. Fararo, ed., *Purpose, Meaning, and Action: Control Systems in Sociology* (New York: Palgrave Macmillan, 2006).

For an introduction to the causes and resolution of conflict, see:

- Richard S. Marken and Timothy A. Carey, *Controlling People: The Paradoxical Nature of Being Human* (Samford Valley: Australian Academic Press, 2015).

For examples of how external and internal contexts interact to produce what makes up the mind, see:

- Batja Mesquita, Lisa Feldman Barrett, and Eliot R. Smith, *The Mind in Context* (New York: Guilford Press, 2010).

An easy-to-read book about priming and other subtle influences on behavior is:

- Adam Alter, *Drunk Tank Pink: And Other Unexpected Forces that Shape How We Think, Feel, and Behave* (New York: Penguin Group, 2013).

A good but scholarly article on priming is:

- John A. Bargh, "What Have We Been Priming All These Years? On the Development, Mechanisms, and Ecology of Nonconscious Social Behavior," *European Journal of Social Psychology* 36 (2006): 147-168.

Endnotes

1. This view of the environment is based on D. Ford, *Humans as Self-Constructing Living Systems,* 51; and Ashby, *Design for a Brain,* 2nd ed., 36. From a Perceptual Control Theory perspective, however, the environment begins from the ends of the nervous system's sensory and motor neurons and includes the body's internal environment. This broader view is not presented here in order to simplify the discussion for general readers by focusing on the body's external environment.

2. Appendix 1 describes five levels: the broader environment, the immediate environment, the person/individual level, the sub-personal level, and the cell and molecular level. See that appendix for details.

3. For example, see Scharfstein, *The Dilemma of Context,* 67-69, 185, 189.

4. Powers' use of the term "environment" is broad and includes all that is outside the nervous system, including what is inside as well as outside a person's body.

5. Such tendencies for people to overlook environmental and situational causes of actions and outcomes in favor of personality traits and individual dispositions is called "the fundamental attribution error." For details see Ross and Nisbett, *The Person and the Situation,* 79.

6. For evidence supporting these statements see Morris and Peng, "Culture and Cause," and Miller, "Culture."

7. Runkel, *People as Living Things,* 4-7.

8. For example, "Two out of three Americans say that everything they need to know to do their job was learned on the job—not through classroom preparation to qualify for those jobs," according to Carnevale, Gainer, and Villet, *Training in America,* 23.

9. Dewey, *Human Nature and Conduct,* 296.

10. Piantadosi, *Biology of Human Survival,* 84-85.

11. Ibid., 164-70.

12. Wener, "Effectiveness"; and Wener, Frazier, and Farbstein, "Three Generations."

13. The offloading of memory onto the world and use of the environment in these ways to do things with less intelligence and effort is called "scaffolding" by Clark, *Being There,* 45-47.

14. Baumeister, Heatherton, and Tice, *Losing Control,* 182, 211-213.

15. Chriss, *Social Control,* 1.

16. McClelland, "Understanding Collective Control Processes, in *Purpose,* 37-38.

17. The issue of compliance is discussed well in the chapter on "Conformity" contained in Aronson, *The Social Animal,* 10th ed., 13-57.

18. McClelland, "The Collective Control of Perceptions," 88.

19. However, when their references or your references are different, conflict can result—this being a topic discussed later in this book.

20. E. Ford, *Love Guaranteed*, 27.

21. Barker, *Ecological Psychology*, 169-78; Wicker, "Behavior Settings Reconsidered," 622.

22. Marcus, "Making the Mind," 141.

23. Reynolds, Gottlieb, and Kastner, "Attention," 1114.

24. Milliman, "Using Background Music"; North, Hargreaves, and McKendrick, "The Influence of In-Store Music," 272; Dijksterhuis and Smith, "The Unconscious Consumer," 198.

25. As William James has pointed out, "A faint tap *per se* is not an interesting sound; it may well escape being discriminated from the general rumor of the world. But when it is a signal, as that of a lover on the window-pane, it will hardly go unperceived"; from James, *The Principles of Psychology*, vol. 1, 417-418.

26. Srull and Wyer, "The Role of Category Accessibility," 1660-1672.

27. Bargh, "What Have We Been Priming," 152.

28. Tipper and Weaver, "Negative Priming," 4317; Lord and Levy, "Moving from Cognition to Action," 351-4; Johnson, Chang, and Lord, "Moving from Cognition to Behavior," 389-393, 401.

29. At the time of this writing, a classic video showing this effect was available at https://search.yahoo.com/search?fr=mcafee&type=C011US977D20151112&p=selective+attention+test. You might give it a try if it is still available.

30. Magistretti, "Brain Energy Metabolism"; also Magistretti, "Low-Cost Travel in Neurons," 1349.

31. To control theorists, sensitization or priming effects seem to represent a change in the gain of concerned neural networks.

32. Dijksterhuis, "Automatic Social Influence," 99.

33. Such nonconscious mimicry of interaction partners is referred to as *the chameleon effect* by Chartrand and Bargh, "The Chameleon Effect."

34. These examples hold for most situations. Of course, exceptions occur, related to one's references. For example, depending on where you work: you might take off your clothes, if you are an exotic dancer; look for the shortest and longest lines, if you are a supermarket manager; or look for motion in a supermarket, if you are a security guard at night.

35. Berger, Meredith, and Wheeler, "Contextual Priming."

36 Harris, Bargh, and Brownell, "Priming Effects of Television."

37 Yi, "Contextual Priming Effects," 8.

38 Sutherland, *Advertising and the Mind,* 3rd rev. ed.

39 Bargh and Gollwitzer, "Environmental Control of Goal-Directed Action," 110; Gollwitzer, Bayer, and McCulloch, "Control of the Unwanted," 485-515. The latter reference indicates that implementation intentions do not work when the goal intention is weak; other factors include strength of commitment to the implementation intention and the strength of the mental link between the if-part and the then-part of the implementation intention (see 491-492 for details).

40 Kirsch and Lynn, "Automaticity in Clinical Psychology," 509. Also, as Richard Marken points out in "Making Inferences About Intention," intentions are another word for a reference state, and intentional behavior is a process of control as explained by control theory.

41 Based on feedback, you adjust what you are doing to better achieve the reference you are seeking to achieve or maintain. That is, your habitual patterns of behavior are not exactly the same each time, but are similar enough to be classified as habits.

42 Baldwin, "Habit, Emotion, and Self-Conscious Action," 36; Ouellette and Wood, "Habit and Intention in Everyday Life," 55-6; Triandis, "Values, Attitudes, and Interpersonal Behavior," 204; Klöckner, Matthies, and Hunecke, "Problems of Operationalizing Habits," 397.

43 Dewey, *Human Nature,* 16.

44 Ibid., 20.

45 Berger, *The Developing Person,* 226.

46 Bohannon, "The Theory," 1616.

47 Ibid.

48 "Drugs: What You Should Know" (Jacksonville, Florida: Nemours Foundation, 1995-2009); accessed October 2, 2009, from http://kidshealth.org/PageManager.jsp?dn=KidsHealth&lic=1&ps=207&cat_id=20140&article_set=22660; "Impulse Control Problems Linked to Antipsychotic, FDA Warns," accessed May 3, 2016 from http://www.medscape.com/viewarticle/862796-print.

49 The term "goals" used here is a synonym for references. The phenomenon of people selecting environments and of environments selecting and rejecting people has been recognized and commented upon by Barker and Wright, *Midwest,* 64, in their discussions of behavior settings. Goal-based situation selection is discussed by Kelley et al., *An Atlas of Interpersonal Situations,* 431-447.

50 Russell, "Core Affect." The type of situation you are in, however, is largely determined by your perception of that situation, an internal influence.

CHAPTER 7

PERCEPTION OF YOUR ENVIRONMENT

The Big Picture

What you perceive around you may not be similar to others' perceptions—even when seeing, hearing, tasting, smelling, or otherwise sensing the same thing! These differences are often obvious among people of different ages and cultures. For example, a clothing style that looks great to a teenager may look terrible to you, and what you think looks great may not look good to someone else. Similarly, a way of behaving that seems appropriate to someone from one culture may look inappropriate to someone from another culture. For example, many Westerners blow their noses into handkerchiefs which they then put into their pockets—a practice that is disgusting to many Asians. Likewise, many Asians, Hispanics, and Africans eat goat, which many Americans do not find appealing. Men and women too may see things differently, as discussions between married couples sometimes indicate (Just ask my wife; she certainly sees more dirt around our house than I do!).

But people do not have to be very different to perceive things differently. Differing perceptions can also hold for people of the same sex, age, ethnic group, culture, religion, and family. Not everyone has the same taste in music. Pop music may be enjoyed by some and annoying to others. Not everyone likes the same food. A dish, such as snails or oysters, may be a delight to one person and revolting to another. Not everyone belongs to the same political party and subscribes to the same ideas about how well a particular politician is doing. Depending on a viewer's political orientation, a politician such as a President or Prime Minister may be perceived either as disastrous for the county or a great and wise leader.

Such differences in perception occur for many reasons. For one thing, the experiences that people have are different. These differing experiences lead to different learning and memories that may then result in differences in perception. For example, a dog may be an object of affection to one person or feared by another, depending on whether the person grew up with friendly dogs in the past or was bitten as a child. The environment and context you are in also affects your perceptions and the perceptions of others. For example, hearing a sound in your house during the day may be perceived differently than hearing the same sound during the middle of the night. Similarly, hearing the word "nine" may be perceived differently depending on whether you are in an English-speaking country or if you are in Germany (where a sound like "nine," spelled "nein" in German, may be understood to mean the word "no" rather than the number "9").[1] Your internal states also affect your perceptions. For instance, a glass of water is perceived differently if you are thirsty or not.[2]

However, as important as the differences are the similarities in perception that different people have. Such similarities are necessary if people are to work together, communicate effectively, and function as a society. When you perceive things the same way as other people around you, you are "on the same page" so to speak, especially if your reference values are similar, as they may be due to similar experiences in education, exposure to media, and interactions with others in your culture. Such similarities enable you to understand other people and to act accordingly. For example, you and others usually perceive behavior settings similarly and tend to act in normal ways within those settings because of common references that you have–such as being quiet in a library or paying for goods acquired in a store or eaten in a restaurant. You also understand the same language and perceive most words the same way as the people with whom you converse. If not, you wouldn't have much of a conversation, would you?

Such similarities and differences are important because people who perceive things differently often act differently. For example, seeing someone positively or negatively, such as a friend or a possible thief walking towards you, affects how you feel and what you do. Thus, you may smile and greet the friend, but do what you can to avoid the possible thief.

In short, how you perceive your environment affects what you do. How others perceive what is around them affects what they do, too. Realizing this will help you to better understand why you do the things you do and why others do what they do. It will also help you understand how and why other people try to manipulate your perception in order to have you do what they want you to do.

What is Your Environment Really Like?

The world as you perceive it, your environment, is a construct of your brain.[3] Colors, sounds, tastes, and smells as well as beautiful and ugly people, good and bad politicians, and the meaning of other objects and events that you sense are all interpretations of what is in your environment, not real aspects of what is around you. The taste of lemonade, for example, only exists in your brain—not in your environment.[4] And, as William Shakespeare wrote in his play *Hamlet*, "There is nothing either good or bad, but thinking makes it so."

Highlight 7.1 — The Zen of Perception

"Thoughts and real things differ in the same way that literary characters differ from living breathing persons. Both are creations of the human mind that have no material existence." "... the entire world as we *know* it is ... an artifact of the mind."

(Radcliff and Radcliff, *Understanding Zen*, 45, 65)

"The belief that our perceptions are precise and direct is an illusion—a perceptual illusion."

(Kandel, *In Search of Memory*, 302)

"Perception should not be viewed as a grasping of an external reality, but rather as the specification of one ...."

(Maturana and Varela, *Autopoiesis and Cognition*, xv)

"The world is constructed. The environment contains no information; the environment is as it is' (von Foerster, 1984, p. 263). The world is virtual information; only an observing system produces factual information. Every bit of information is an internal construct (von Foerster, 1999)."

(Vanderstraeten, "Observing Systems," 302)

Why is this? It is partly a result of how your nervous system and brain are organized and operate. For instance, objects don't inherently have color. There is no color "out there." Color is something that your brain constructs from the signals that it receives from the sensory cells in your eyes, which are sensitive to the light waves of photons that they detect. Similarly, food and spices don't have good and bad taste. The taste that you detect is simply an interpretation of signals received from your taste buds—signals that have been activated by molecules to which your taste buds are sensitive. Sounds and music also don't exist outside yourself, although variations of air pressure do. Such variations are sensed by cells in your ear and converted into neural signals that you then mentally experience as sound. There is no heat and cold "out there," only moving molecules with more or less kinetic energy.[5] In short, all of the sensations, feelings, and other perceptions you experience, as well as the concepts you have (such as good and bad, animals and plants, countries and planets) are mental creations of your brain—constructions of what you perceive to be reality.[6] And many of these creations are affected by the culture in which you grew up. For example, whether you perceive a dog as something to be eaten or not depends on your cultural upbringing.

Highlight 7.2 Illusion: An Example of How the Brain Constructs Reality

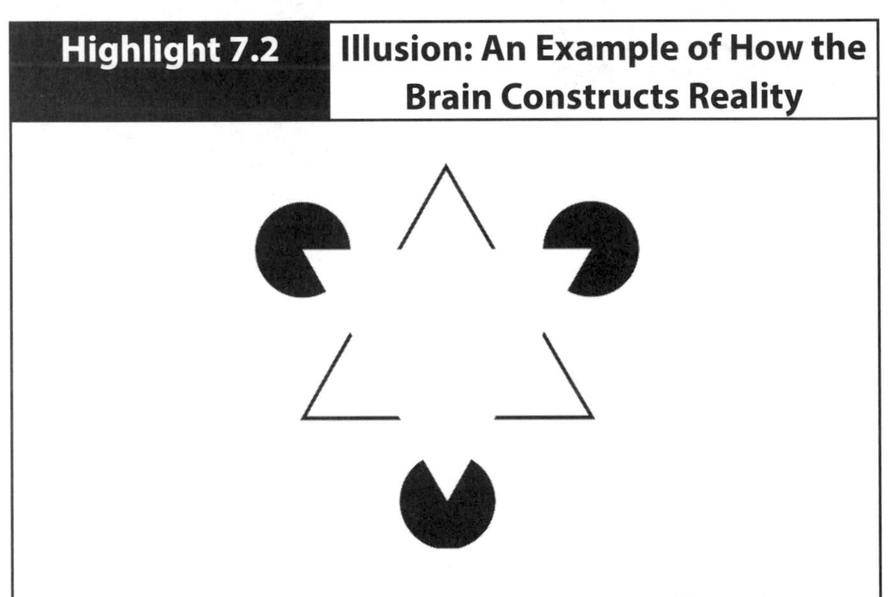

The brain creates shapes from incomplete data—for example, filling in the missing lines of a triangle. The triangle does not exist physically, but is created by your brain. If you hide some parts of the picture, the triangle vanishes.

Note: This illusion is called the Kanizsa triangle and was first described by Gaetano Kanizsa in 1955.

Even something as basic as a smile or a gesture can be perceived differently. In Japan, a smile is a common indicator of discomfort or embarrassment[7] whereas in Western countries, a smile is usually interpreted as a sign of happiness or amusement. The circular hand gesture of the thumb touching the end of an adjacent finger is perceived by most Americans as meaning "OK"; "money" to the Japanese; "zero" or "nothing" to the French; an insult to Turks and Brazilians; and a threat signal or obscenity to many Arabs. The "thumbs-up signal" may mean you are hitchhiking; indicating agreement or "OK"; showing the number "1" or "5"; implying that someone is "homosexual"; or giving an insult such as "up yours" or "sit on this."[8]

Highlight 7.2 is an example of how our brains construct reality. More illusions can be seen online at http://en.wikipedia.org/wiki/optical_illusion.

Knowing that what you perceive is a construct of your brain may help you realize that different people often perceive reality differently. As a result of their differing perceptions, they may act differently. Furthermore, you can change how you view the world, to some extent at least—a realization that if acted upon, can change your life for the better.[9]

What is Perception?

According to William Powers, "perception" refers to the entire series of events, following stimulation, that occur in the input part of the brain—from sensory receptors to the highest centers of your cerebral cortex.[10] Some perceptions are conscious; we are aware of them. Most are not. In Powers' model (discussed in Chapter 3), he calls neural currents related to perceptions "perceptual signals." Such signals activate behavior when they result in error signals due to important differences between those perceptual signals and related references such as your goals and homeostatic set points. In other words, as Powers puts it, "Behavior is the process by which we act on the world to control perceptions that matter to us."[11] See Highlight 7.3.

Highlight 7.3 Control Systems

"Control systems, or organisms, control what they sense."

(Powers, *Behavior: The Control of Perception*, 355)

As mentioned in Chapter 3, Powers distinguishes among various types of perceptual control systems. For example, some systems control perceptions of intensity signals from your sensors and resulting sensations such as the pain that results from touching a hot surface or pleasant feelings of sexual activity. Other systems control perceptions of what he calls "programs" where you face a series of choices or alternatives about what to do, such as when driving a car to work or eating at a restaurant. Such control systems act to match your perceptions of your external world and of your body to your references—thereby enabling you to achieve many of your goals and continue living. The different levels and types of perceptions you control, as suggested by Powers, are shown in Highlight 7.4.[12] These range from "intensity," the simplest level, on up to the complex "systems concept" level.

Highlight 7.4 Levels of Perception and Control

1. **Intensity:** Neural impulses that result from the conversion of external energy in sensors related to the sensing of light, sound, pressure, vibration, deep touch, balance, taste, chemicals in the air, tissue destruction, and deformity. The rate of neural impulses gives you a sense of the outside world and how much energy is affecting your sense organs.
2. **Sensation:** perceptions of color, taste, sound, pain, balance, and touch.
3. **Configuration:** perception of patterns in sensations that result in perception of your body's position and nameless forms, objects, and things.
4. **Transition:** perceptions of change, such as turning, rising, dropping, expanding, shrinking, straightening, flowing, rolling, rotating, twisting, increasing, decreasing, and sounds growing louder and softer.
5. **Event:** Occurrences that have a beginning, middle, and an end, such as opening a door, sitting down, an explosion, a slap, an opera performance, and a football game.
6. **Relationship:** Spatial ordering such as in front of, behind, and beside; temporal ordering, such as before, starting, during, ending, and after; other associations such as accompanying, confronting, helping, cheering, teaching, and causation.
7. **Category:** Words indicating a classification, such as group, set, kind, sort, type, denomination, collection, color, grade, and clan. Examples include males, females, stone, grass, democracy, socialism, personality, intelligence, and corporation.
8. **Sequence:** Things perceived to occur in an order, such as the organization of a cooking recipe, singing a song, and checking the oil level in a car.
9. **Program:** A program contains alternative sequences and branching points, such that *if* this is perceived to be the case, *then* you will do that. If you drive to and from work, you have programs for dealing with perceptions of road conditions, ranging from unconsciously moving the steering wheel, brake, and gas pedal to more consciously choosing which route to take if there are traffic delays or errands to run.

10. **Principle:** Such perceptions include those of moral matters such as honesty, keeping children safe, behaving according to your values (such as "be considerate to others"), and heuristics (such as "control the center of the board in chess").

11. **System Concept:** These perceptions concern the way things hang together such as they do when you perceive music, the science of astronomy, and organizations such as the Boy Scouts, the Catholic Church, and the criminal justice system.

Sources: Runkel, *Casting Nets*, 198-210; Powers, *Making Sense of Behavior*, 135-52.

Other descriptions of perception are different from that of Powers, but most start with the senses and end with the assignment of meaning to what is sensed and perceived. For example, a definition given in Highlight 3.2 indicates that perception is "The process or result of becoming aware of objects, relationships, and events by means of the senses, which includes such activities as recognizing, observing, and discriminating." Similarly, Alvan Noë, a professor of philosophy, writes that, "To perceive is not merely to have sensory stimulation. It is to have sensory stimulation that one understands."[13] These ideas emphasize a conscious awareness of perception, in contrast to other views, including Powers', that indicate many perceptions are unconscious. For example, your body's continuous yet usually unconscious perception of its position and stability helps keep you from falling down when standing and from falling off chairs when sitting.

The view taken here is that perception can be either conscious or unconscious, since so much of what we do occurs without consciously perceiving what is around us. Perception also occurs continuously by means of all our senses and is what we control when we behave. In addition, conscious perception is viewed as a construction within the brain of the environment being sensed—an individual's interpretation of what the world is like, based on the person's neural structures that are a result, in part, of the person's unique experiences in the world.[14] As indicated in "The Big Picture" at the beginning of this chapter and to be discussed more, these constructed perceptions often differ among people, even when the same objects and events are viewed or otherwise sensed.

The Importance of Your Perceived Environment

Your ability to perceive the environment around you is important for many reasons. In particular, your perceptions

- **are essential to your survival**. Your perceptions are needed to determine what you should avoid, approach, and make use of. They also help establish when you should take action and which actions that you take are effective in maintaining your autopoietic being. For example, your ability to perceive food enables you to obtain nourishment. Your perception of possible hazards such as a hot stove, spoiled food, or a car moving towards you, helps you avoid being harmed. Your ability to perceive weather conditions enables you to dress properly and maintain an optimal body temperature. Perceptions of risk help you to take precautions.
- **evoke positive and negative feelings** about people, things, and events observed, felt, tasted, smelled, or otherwise experienced. These feelings seem to be a reflection of the error signals that are occurring and help to guide what you do.[15] They are a way of determining, quickly and broadly, what is good for you and what is not. Good or pleasant feelings usually indicate those things and actions that may help you to survive and maintain your autopoietic being. Bad or unpleasant feelings usually indicate things and actions that may act against your survival, disrupt your body's homeostatic states, and possibly lead to your death.
- **indicate resources and opportunities** available to you for achieving your goals and other reference conditions, including things essential to your survival, such as food, water, shelter, clothing, support from others, and other environmental affordances. "Affordances" are aspects of the environment that you recognize as providing occasions for use and action. For instance, a bed may afford or provide you an opportunity to sleep, and a restaurant may provide an opportunity to eat. In other words, there is tight perceptual attunement between you and your environment. Because of this, you are sensitive to features of the world that afford you opportunities for action needed to achieve or maintain your desired perceptions.[16]
- **indicate constraints** that may prevent you from achieving your goals, including obstacles that need to be overcome in order to achieve the

results you need or desire. You may, for example, perceive that you don't have enough money to buy an item you need, or perceive that you don't have enough education to obtain the job you want—realizations that may help you to either do things to overcome those constraints or to revise your goals.

- **stimulate related references.** Perceiving your environment may unconsciously prime or consciously remind you of references you have. For example, seeing a policeman or police car may remind you that you don't want to receive a ticket or be arrested, with the result that, if you are driving a car too fast, you may slow down to the legal speed limit. Similarly, going into a library you may be automatically stimulated to talk more quietly to a companion, after your reference to "be quiet in a library" is activated.
- **evoke the roles you should assume and the norms that you are expected to follow.** Your perceptions also prime and affect what you do in different behavior settings and when interacting with other people. For example, if you are in a classroom, you may assume the role of a student, or a teacher, or even a parent (such as during a parent-teacher conference). Related norms to be followed may be indicated not only by your role, but also by the setting itself, by means of signs, symbols, or directions, such as "No Smoking," by means of feedback provided by others there—smiles, disapproving frowns, threats, or shouts—and by the behavior of others that you may imitate.
- **indicate how well you are doing** in achieving your goals, standards, desires, and other references. In other words, your perceptions inform you of the status of what you are trying to achieve and control. With such information, you are able to adjust what you are doing and change your behavior and environment to better achieve your goals and other references. If resulting perceptions are not what you want, you know that further action is needed.[17]
- **help you learn** what to do in given situations. For example, perceptions of the consequences of your behaviors indicate how beneficial, useless, or injurious they are in given situations, thereby enabling you to react appropriately to similar situations in the future. In fact, nearly all of the reasons given above about the importance of perception are related to

your ability to learn. That is, your perception of the environment either helps you to learn what to do by indicating how well you are doing, or is a result of learned associations—such as the feelings you have about people, objects, and events, as well as your ability to recognize resources, opportunities, and constraints.

To sum up, perception of your environment is important and crucial to your survival. Your perceptions permit you to achieve and maintain your goals and other reference conditions with the result that you live more comfortably, enjoy life more, have fewer problems and difficulties, and avoid unpleasant situations, injury, and death.

What Affects How You Perceive Your Environment?

Many factors influence your perception. Internally, these concern how your body is structured, organized, and functioning. Externally, the physical world affects your perception by the way it activates your sensory neurons, interacts with your body, and affects how your body is organized and operates. Let me explain.

Your Past History

As discussed in Chapter 4, your past experiences have affected your neural structures, learning, and behavior. Those experiences have also affected your perceptions. Research by Vanderijt and Plooij indicates that from the time you were born, your perceptions have developed in somewhat predictable ways. From birth, you had at least some perception of sensations such as light, sound, smell, touch, and pain. But, those perceptions were mixed up and blurred. By about eight weeks of age, you had developed awareness of some configurations, such as a person or something hanging above your crib. At 37 weeks, you began to recognize categories of things like "sweet" and "horse." Similarly, other kinds of perceptions were developed and learned over time, as indicated by Highlight 7.5. Development of those perceptions depended on the experiences you had with your environment. For example, if you had not been exposed to pictures, words, or real horses, you would not be able to perceive the category of objects called "horse" when seen.

Highlight 7.5	Perceptual Development of Babies
Age	**Perception Being Developed**
5 Weeks	Alertness of changing sensations, internal and external
8 Weeks	Recognition of simple patterns, recurring shapes, and structures
12 Weeks	Smooth transitions of sensations, including eye and head movements
19 Weeks	Events – short familiar sequences, including movements and sounds
26 Weeks	Relationships between different things such as distance, inside, outside, next to; between sounds and objects; and flipping a switch to turn music on
37 Weeks	Categories – recognizing that things are similar and belong together
46 Weeks	Sequences, such as constructing and putting things together in a certain way each time; doing things in a certain order to reach a goal
55 Weeks	Programs – things can be done more flexibly in reaching a goal such as setting the table, getting dressed, and eating lunch

Note: The ages shown are averages. Development of perceptions indicated may occur before or afterwards.

Source: Vanderijt and Plooij, *The Wonder Weeks*.

From infancy onwards, your social environment has affected how you perceive things. Other peoples' environments have affected them too. For example, as mentioned before, a smile may be perceived as an indication of discomfort or embarrassment to a Japanese or as a sign of happiness or amusement to a Westerner. Or, as Highlight 7.6 indicates, a traditional Cree hunter and a more Westernized hunter may perceive the same thing very differently. Why these differences? It's because you and others have learned to behave and associate different things as a result of your environment and the experiences that you had.

Highlight 7.6 An Example of Differing Perceptions

A reindeer when becoming aware of someone's presence, instead of running away, stands still, turns its head, and looks directly at the person's face. A Western hunter may perceive the animal as "stupid" for doing this; a biologist may perceive what is happening as an adaptation to predation by wolves. However, a Cree hunter of northeastern Canada, perceives the animal as offering itself to the hunter, "quite intentionally and in a spirit of goodwill or even love toward the hunter."

(Tim Ingold, *Perception of the Environment*, 13)

Similarly, eclipses of the sun and moon were once perceived as dire predictions about the world and the actions of gods. Astrologers still give such occurrences special meaning. However, many people now perceive eclipses as nothing more than an interesting sight based on the natural movement of the earth and moon around the sun. It all depends on what you have learned and believe. As Richard L. Gregory wrote in his book, *Seeing Through Illusions*, "We *see* phenomena as we *understand* them by the current model in mind."[18]

In addition, throughout your life, you learned to link and associate certain feelings with your perceptions. As a result, if you perceive an object or event now, you have either a good, bad, or neutral feeling associated with that perception. The feeling depends on the previous associations formed, if any.

Such associations may cause you to expect that certain types of perceptions will occur when you experience certain phenomena. These expectancies may bias your perception so much that you actually perceive what you expect to perceive. For example, if you see a politician that you like on TV, you may perceive what he or she is saying is reasonable, whereas if you dislike that politician, you may perceive what is being said as nonsense. Such expectations can result in what is called the "placebo effect." This effect often occurs when someone expects or believes that something will have an effect and, because of that belief or expectation, the effect is perceived to occur and may even do so—even though the presumed cause is inert and has no inherent power to produce what is expected. An example is when a sham miracle cure or treatment, such as a detox footbath, results in perceptions that one's condition has improved, or when an expensive brand-name drug is perceived to be more effective than a less expensive but identical generic drug.[19]

In short, the experiences you have had with your environment influence how you perceive things.

Your Present Condition and Capabilities

Your current goals, needs, feelings, knowledge, skills, and the present condition of your body also affect your perception. Many of these factors are a result of your past history as discussed in the previous section but are mentioned here in order to focus more on you at the present time. Here are some examples:

- **Your goals:** If you are actively doing things to achieve a goal, your attention and perception are more attuned to things and events related to that goal. You also are less sensitive to things and events that are not related to achieving that goal. For example, a man may barely listen to his wife when watching a TV football game he is interested in.[20] More generally, however, even when you are not actively pursuing a goal or other reference, your references can influence your perception. For example, you may become aware of deviations from a reference you have, such as seeing someone wearing what you consider inappropriate clothing at church, work, or school. Or you may avoid perceiving things that you do not agree with, such as a TV program, person, book, or explanation that differs from your goals and values.
- **Your expectations** also influence your perception, especially in ambiguous situations. Those expectations prime you towards perceiving what you expect to perceive. And so, you are inclined to experience what you expect to experience—ranging from judging the merits of statements made by politicians or radio commentators you like or dislike, to possibly judging others, such as members of minority groups, in biased and stereotyped ways. Such priming and related perceptions also lead to self-fulfilling prophecies. For example, teacher expectations about individual students typically result in those students being perceived in a certain way and dealt with accordingly: as a troublemaker, cooperative, intelligent, stupid, or a leader.[21] People's expectations lead to idioms such as "seeing the world through rose-colored glasses" and placebo effects such as feeling less pain after taking a fake pain medication that is actually an inert sugar pill. Other examples of the effects of expectations in action include Ouija board spelling, dowsing, and hypnotic effects as well as therapy sessions where clients are asked to imagine likely outcomes, thereby priming them to achieve those outcomes.[22]

- **Your moods and feelings** also affect what you perceive. When you are in a good mood, you often perceive things more positively. When in a bad mood, your perceptions may be more negative.[23] When you are hungry, you perceive food-related items more easily. When you have feelings of thirst, thirst–quenching items are perceived more readily.
- **Your knowledge and skills:** Without certain knowledge and skills, you may not even perceive certain things. For example, a musician can perceive and identify musical notes such a C and C sharp. Can you? How about a forester, who can perceive and identify different types of trees, and whether they are healthy or not. Can you? The same with doctors and health-related matters—which is a reason why you see a doctor when you are ill, isn't it? You also know what is normal behavior and have norms concerning such behavior. Your knowledge of what is normal helps you to perceive the appropriateness and inappropriateness of events in your environment. What you know also affects how you perceive the meaning of sounds you hear. For example, does a word that sounds like "mama" mean "mother" as in English or "mother's brother" as in the Nepali language?
- **Your physiology and the physical state of your body** also affect your perception. For example, color blindness affects the perception of color. Deafness affects the perception of sound. Temporary bodily changes such as levels of hormones and neurotransmitters can increase or decrease your sensory thresholds and, therefore, your perception and resulting behavior.[24] For instance, the release of endorphins by your body is an important factor in the perception of pleasure and can cancel or reduce your perception of pain because they have effects similar to morphine.[25]

Your Present Environment

You continuously perceive your environment, and your environment continuously influences what you perceive and do. Sometimes these influences are more noticeable, such as events occurring that are relatively sudden, novel, or unique. Many occur naturally, such as gravity, and day and night light cycles. Others occur as a result of sensations that you experience as you move about in the world around you.

However, perception of your environment is not one-way. You don't simply perceive the world around you. People in your environment also try to affect your perception. They often try to influence and even manipulate what you perceive in order to affect what you do. They may do so either intuitively and unconsciously, or consciously, fully realizing what they are doing. Such influencers and manipulators include family members, friends, teachers, co-workers, merchants, and people you may have never met such as advertisers, politicians, religious leaders, and con artists.

Why do people do this? Why do they try to influence what you perceive and do? They do so in order to achieve their own goals and other references—some of which they are not even consciously aware of. Since your behavior is so closely linked to your perception, they try to affect what you perceive so you behave in ways that help them achieve their goals. Such goals could be profits, as in the case of a businessman; the enjoyment of being with you, like a friend or lover; to obtain your vote, as a politician; or perhaps to affect how well you think of them, such as a college teacher seeking good evaluations at the end of a course.

We discussed some ways that others try to influence you in Chapter 4. However, many other techniques exist, such as:

- **Directing your attention**, as by means of the intensity, movement, novelty, troubling nature, or sexual content of things you perceive—as occurs on TV, in movies, in magazines, and online. Exceptions to what you continually perceive and things that are different, meaningful, or attractive also tend to capture your attention. That is why some news programs on TV have continually moving backgrounds, and why automobile advertisements often include beautiful women and handsome men.

- **Using environmental props.** Some objects and arrangements, sometimes called "environmental props," can have a great effect on how you perceive people, things, and events. These include weapons such as a gun or knife; clothing such as uniforms; the positioning of furniture, such as chairs in an office, face-to-face across a desk or side-by-side; and the display of objects such as awards, diplomas, and pictures showing the status, position, or interests of someone.[26] As Dan Ariely points out in his book *Predictably Irrational,* even the artful presentation of food on a nice dish or of wine in a nice wineglass can make a difference in how good it tastes.[27]

- **Framing and spin techniques** are also used by others to have you perceive and think in certain ways. For example, when talking about immigrants, a politician might use either the words "undocumented immigrants" or "illegal aliens" depending on how he or she wants you to think. Each usage of words primes you in different ways, with "illegal aliens" usually resulting in more negative perceptions. Similarly, for example, one side will frame their talk by referring to "The Affordable Health Care Act" while the other side will use the term "Obamacare;" one side may say "terrorist" and the other side "freedom fighter" or "martyr;" and "the estate tax" may be referred to as "the death tax" by those who oppose it. "Spin" is a more manipulative use of a frame, often when something embarrassing has happened, to make the occurrence sound normal or good.[28] For example, instead of mentioning that civilians were killed by soldiers, you may hear that some "collateral damage" occurred. See Highlight 7.7 for more about frames and framing.

Highlight 7.7 Framing

"**frame of reference**... a set or system (as of facts or ideas) serving to orient or give particular meaning: VIEWPOINT, THEORY."

(Mish, *Webster's Ninth New Collegiate Dictionary*, 489)

"Frames are mental structures that shape the way we see the world. As a result, they shape the goals we seek, the plans we make, the way we act, and what counts as good or bad outcome of our actions."

"It is a general finding about frames that if a strongly held frame doesn't fit the facts, the facts will be ignored and the frame will be kept."

(George Lakoff, *Don't Think of an Elephant*, xv, 37)

- **Building rapport** by expressing genuine interest in you, dressing appropriately, letting you do most of the talking, listening well, reflecting empathy with you, matching your body language, and matching your vocal tone, speech pattern, and breathing rate.[29]
- **Advertising and marketing** that is aimed at shaping perceptions of products, services, organizations, and people. In advertising, politics, business, and life, "perception is the reality" and so, according to two marketing specialists, Al Ries and Jack Trout, "Truth is irrelevant. What

matters are the perceptions that exist in the mind."[30] Similarly, when Wesley W. Zimmerman writes about buying, marketing, and customer care, he states, "We make decisions and act upon them based on perceptions."[31] That is a reason why advertisers and marketers spend so much time, money, and effort trying to affect our perceptions, rather than just giving us the facts.

- **Social marketing and cause marketing**, which seek to change personal behavior or public policy, usually for the better, by using advertising and marketing techniques. Examples include persuading people to avoid or stop risky practices such as smoking, drug abuse, unprotected sex, littering, and being careless with campfires; seeking counseling for compulsive gambling or spousal abuse; getting vaccinated; reducing high cholesterol levels; using a safety belt; joining or giving financial support to groups that benefit society; or otherwise doing things to improve health, prevent injuries, protect the environment, and benefit communities.[32]
- **Formal education** at schools, colleges, and universities develop your ability to perceive everything from letters, words, and symbols to other aspects of the world around you including cause and effect relationships; the interpretation of graphs and maps; categories of animals and plants; sports values such as playing according to the rules; and doing things "on time."
- **Training and development** at workplaces including induction, orientation, staff development programs, and apprenticeships. These seek to have you view the organization and your work in certain ways. For example, if you are hired to work in an organization, you learn to perceive who your boss is, don't you?
- **Perception management** as used by governments, businesses, other organizations, and individuals including politicians. The intent is to affect perceptions of the image, identity, or reputation of an organization, person, or product. For instance, organizations and politicians may use this approach after scandals, accidents, product failures, and poor ratings by industry groups; after good news and positive events, including those that may be because of luck; during contentious negotiations; and before events that may be viewed negatively by people, such as the opening of a nuclear reactor facility.[33]

- **Propaganda**, which often overlaps with perception management, framing, and spin techniques. Propaganda simply asserts something as a fact such as "Obama is Muslim"; gives an unfair advantage to one point of view; uses vague words and generalities like freedom, security, and change; and uses false choices such as "You are either with us or against us."[34]
- **Drugs and other medical techniques** including antipsychotic drugs and fluoxetine (Prozac) that can affect perception in ways that include the reduction of hallucinations, delusions, depression, pain, and anxiety.[35]

In addition to the active use of techniques by others to influence your perception, sometimes your environment influences what you perceive more indirectly. Here are a few examples:

- **Goal activation:** Sometimes a goal or other reference is activated when you encounter a familiar situation or interact with others. The goal, norm, program, or other reference that is primed or activated will then affect your perception of the situation being encountered.[36]
- **Mere exposure:** The more we are exposed to something and the more familiar it is, the more we tend to like it if the object, person, or event has caused us no discomfort or harm. Seeing something and liking it is a perception. Such liking of neutral objects, persons, and events occurs naturally. Because we often like things that are familiar to us is why you may like a new song more after you have heard it several times.[37]
- **The context:** Different contexts can influence your perception, even when you experience the same sensations—as on a roller coaster compared to a bumpy airplane ride, or hearing a sound in the house during the day compared to the middle of the night.[38] Such contextual influence occurs constantly and naturally and is a type of priming and framing. In other words, our perceptions and thinking are constantly being primed and framed by the context that we are in at the moment.
- **The nearness of desired perception attainment:** The timing of desired perception attainment may affect how you perceive things. That is, desired sensations, goals, and consequences whose attainment is nearer often seem more desirable than those that may be achieved at a

> later time. That is why you and others often give into temptations, such as eating a banned food when on a diet or buying a tempting item of clothing or electronics being looked at rather than keeping to a budget or reducing long-term credit card debt. In short, using popular language, we often perceive nearer rewards as greater than more distant ones.[39]

In ways such as these, your environment affects your perceptions, often without you even knowing that you are being influenced. See Highlight 7.8 for some more important thoughts about perception.

Highlight 7.8 **Some Important Thoughts About Perception**

"Perception depends on attention and expectations—for instance, what you're expecting to get.… Perception also depends on a lifetime of prior experience."

(Wang, *Neuroscience*, 26)

"Human perception is skewed toward the features of the world that matter with respect to human needs and interests."

A. Clark, *Being There*, 25)

"It is not the actual environmental situation that leads to responses, but that situation as perceived by the organism."

(Powers, *Behavior: The Control of Perception*, 48)

"Social behavior and communication requires that there must be common perceptions (meanings)…."

(Burke, "Perceptions of Leadership," 287)

Why Do Different People Perceive and Behave Differently?

To the extent that you share common environments and experiences with others, your perceptions will be somewhat similar. Thus, people from the same culture will often perceive things similarly. So will people from the same family, peer group, working group, or other subculture. In fact, common perceptions and related references are needed if social groups and society are to function well. At least some perceptions held in common are involved in social activities such as conversations, meetings, games, and religious rituals; social designations and social identities (mother, father, teacher, criminal, ...); and perceptions about money, norms, social roles, fashions, language, and traditions.[40]

However, people's perceptions may also differ because of differing life histories, experiences, conditions being faced, and capabilities. These differences in perception may cause others to act in the world differently than you do. Recall our earlier example—if you perceive a dog as friendly and someone else perceives it as something to be feared, you will both respond differently to the dog.

However, besides differences in perception of our environments, at least four other important factors affect what we do. As mentioned before, these are:

- Our goals, values, and other references
- Our capabilities, including our knowledge, skills, and physical abilities
- The environments, contexts, and disturbances that we face
- The feedback that we receive.

Differences in one or more of these factors account for why you, I, and other people behave differently even when we are in the same environment.

Preview of the Next Chapter

Now that you have an idea of what your immediate environment refers to, how you perceive it, and how it tends to influence your perceptions and behavior, we will move on to Chapter 8 and a discussion of your broader environment or macro-environment as it is called by some. This refers to more distant aspects of the world that influence what you do. And so, we'll discuss how your cultural context, economic conditions, the mass media, the medical system, technology and science, organizations such as governments and corporations, and even Mother Nature on the far side of the world, influence what you and others do.

Further Reading

An interesting book about perception and how we interpret what we see is:

- Richard Gregory, *Seeing Through Illusions* (Oxford: Oxford University Press, 2009).

A related book filled with illusions is:

- Al Seckel, *Optical Illusions: The Science of Visual Perception* (Buffalo: Firefly Books, 2006).

An academic introduction to perception of the environment by a social anthropologist is:

- Tim Ingold, *The Perception of the Environment: Essays on Livelihood, Dwelling and Skill* (London: Routledge, 2000).

For an easy to read book on how organizations influence our perceptions of them, their products, and services, see:

- Al Ries and Jack Trout, *Positioning: The Battle for Your Mind* (New York: McGraw-Hill, 2001).

Endnotes

1. Noë, *Action in Perception*, 32.

2. Riehle, "Neuronal Correlates," 487.

3. Some people would use the word "mind" here rather than "brain." I avoid use of the word mind as much as possible, since, presumably, the mind is simply a product of the brain, and using the word mind may conjure up the idea that the mind is a real thing rather than simply a convenient abstraction—a word that denotes something that has no existence in reality other than the abstraction that it is.

4. This lemonade example has been mentioned several times by Richard Marken on the Control Systems Group listserv via csgnet@listserv.illinois.edu

5. von Foerster, *Understanding Understanding*, 215.

6. Kandel, Schwartz, and Jessell, *Principles of Neural Science*, 4th ed., 412; Mingers, *Self-Producing Systems*, 210-211; Maturana and Varela, *Autopoiesis and Cognition*, xv.

7. Committee on Opportunities in Basic Research in the Behavioral and Social Sciences for the U.S. Military, "Nonverbal Communication," 50.

8. Pease and Pease, *Body Language*, 118-120.

9. One way to do this is suggested by Ford in his book *Love Guaranteed*. This book, built upon William Powers' Perceptual Control Theory, describes what the author considers to be a "guaranteed" system for improving a marriage.

10. Powers, *Behavior: The Control of Perception*, 2nd ed., 34.

11. Powers, *Making Sense of Behavior*, 17.

12. Powers writes, "Don't take these levels I propose too seriously." "I think of them as a useful starting-point for talking about the hierarchy of control; they'll do until something better comes along," *Making Sense of Behavior*, 135.

13. Noë, *Action in Perception*, 181.

14. This view is in keeping with that of autopoietic theory as described by Mingers, *Self-Producing Systems*, 210-211; Maturana and Varela, *Autopoiesis*, 1980, xv; and Maturana and Varela, *The Tree of Knowledge*, 22-23.

15. As indicated before, Powers considers feelings and emotions to arise from error signals that result from a difference between perceptions and references. For details see "Chapter 17: Emotion," in Powers, *Behavior: The Control of Perception*, 2nd ed.

16. Noë, *Action in Perception*, 21; Gibson, *The Ecological Approach*.

17. Runkel, *People as Living Things*, 21.

18. R. Gregory, *Seeing Through Illusions*, 22. On a more academic level, Marken points out how behavior is seen through the theoretical preferences of an observer, in his article, "Looking at Behavior."

19. For more details see Goldacre, *Bad Science*.

20. D. Ford, *Humans as Self-Constructing Living Systems*, 248, 254-256; Lord and Levy, "Moving from Cognition to Action," 350-356.

21. Tauber, "Good or Bad."

22. Kirsch, *How Expectancies Shape Experience*; Kirsch and Lynn, "Automaticity in Clinical Psychology."

23. Forgas and Bower, "Mood Effects"; Forgas, "The Affect Infusion Model."

24. Becker et al., *Behavioral Endocrinology*, 2nd ed., 325-326.

25. Damasio, *Descartes' Error*, 263; Kandel, et al., *Principles of Neural Science*, 5th ed., 548-552.

26. Forgas, *Social Episodes*, 157-159.

27. Ariely, *Predictably Irrational*, 165.

28. Lakoff, *Don't Think of an Elephant*, 100.

29. Hadnagy, *Social Engineering*, 162-171.

30. Ries and Trout, *Positioning,* 9-10.

31. Zimmerman, *Perception of a Difference,* 2.

32. Earle, *Art of Cause Marketing,* 3; Kotler and Lee, *Social Marketing,* 7.

33. For example, see Elsbach, *Organizational Perception Management.*

34. Good introductions to this technique are Shabo, *Techniques of Propaganda,* and Ellul, *Propaganda.*

35. Grilly and Salamone, *Drugs, Brain, and Behavior.*

36. Bargh, "Auto-Motives," 100-101.

37. Zajonc, "Feeling and Thinking," 31-58; Bohner and Wänke, *Attitudes and Attitude Change,* 76-78.

38. Langer, *Mindfulness,* 38-39.

39. Ainslie calls this tendency "hyperbolic discounting" in his book, *Breakdown of Will.*

40. Magnusson, "Wanted: A Psychology of Situations," 23, 28; McClelland, "The Collective Control of Perceptions," revised draft of a paper.

CHAPTER 8

YOUR BROADER ENVIRONMENT AND ITS INFLUENCE

The Big Picture

Your broader environment consists of the world that lies beyond what you directly perceive and experience. Popular terms used to describe this more distant environment include your nationality, geographic location, national economic conditions, political system, government, religion, organizations, technology and science, historic and current events, and happenings in other parts of the world.

Your broader environment affects you in many ways that are often not apparent. It does this by influencing your immediate environment and affecting what you experience, learn, do, and can do. For example, the language you speak, the way you dress, what you eat, the TV programs you watch, the wages you earn, and whether you are in jail or not are all affected by what lies beyond the immediate environment that you directly experience. That is, you speak your mother tongue because your mother and others learned this language from other people and then used it with you. If the official language of the nation you live in is different, you probably learned that language too, due to policies about languages to be taught at school. You dress in certain ways, such as by using cloth rather than animal skins, because that is what is done in the culture and technological age of which you are a part. People unknown to you grow, process, and transport the food that you buy and eat. People you have never met fund, develop, and produce the TV programs that you watch. If you work for wages, labor laws and economic conditions affect your salary and even whether or not you have a job. You are in jail or not, in part, because of laws passed by others.

In addition, people in your broader environment have designed and developed things to help achieve their goals and, sometimes, the goals of others—for example, by building roads, subways, street signs, cars, and schools; inventing useful things such as money, computers, televisions, air conditioners, and light bulbs; and by discovering and developing things such as vaccines and medical treatments. These influence your behavior. Other people that you have never met, such as politicians, religious leaders, and business persons also try to control or otherwise direct your behavior in ways that they feel is appropriate. For instance, politicians make laws that affect what you and others do. Religious leaders attempt to have you and others behave according to their values and rules of morality. Business people, including boards of directors and managers, try to have you buy their goods and services. Terrorists may seek to have you live in fear. Even though you may never have direct contact with them, they still may influence what you do.

What is Your Broader Environment?

As discussed in Chapter 6, your "immediate environment" is the environment that you directly experience—the people, places, things, and events that directly affect your senses and other aspects of your body, both perceived and not perceived by you (such as gamma rays and some toxic chemicals) but which affect you nonetheless. Your "broader environment" consists of those things that affect your immediate environment but that you do not directly sense, experience, or contact.[1] For example, people in your immediate environment such as your mother, relatives, friends, and teachers, were affected by the behaviors of their parents, the schooling they received, the culture in which they were raised, and their religious upbringing. These factors that affected them can be thought of as parts of your broader environment, at least in a historic sense. However, your broader environment isn't just historic. It also consists of conditions and events occurring now, such as the present state of the economy, wars in other places, government laws and policies, and your family's workplaces. These distant things may affect the behavior of those with whom you interact or the availability of resources, opportunities, and constraints. For example, in the past, the state of the economy may have affected whether or not your parents had a job and if they had to hold down two or more jobs to earn enough money for food and lodging. In turn, that may have affected how they interacted with you—whether they spent much time with you or perhaps whether they were stressed and irritable or drunk, abusive and neglectful.

Highlight 8.1 indicates some of these kinds of relationships between your broader environment, your immediate environment, and you.

Highlight 8.1 Your Immediate and Broader Environments

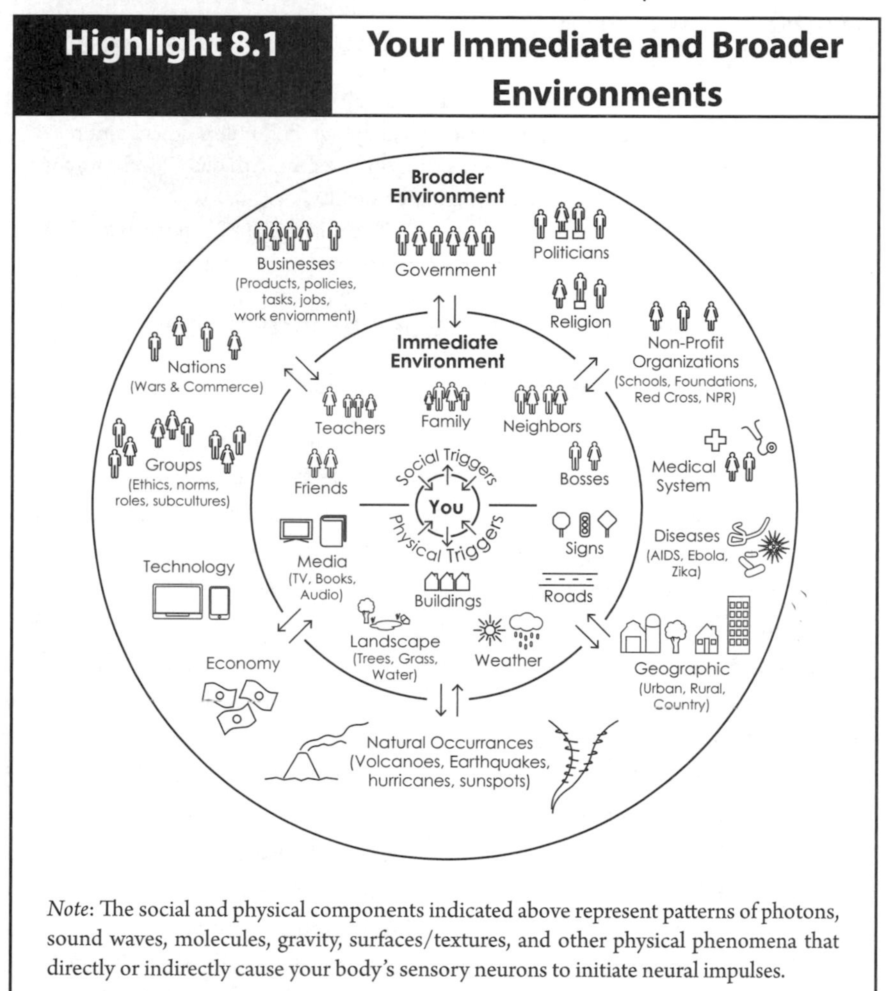

Note: The social and physical components indicated above represent patterns of photons, sound waves, molecules, gravity, surfaces/textures, and other physical phenomena that directly or indirectly cause your body's sensory neurons to initiate neural impulses.

In short, your broader environment is that big world, both past and present, that you do not directly experience, but that affects your life in many ways. The same holds for other people. Their broader environments are also not experienced directly by them, but each person is affected. It is important to understand this idea, because if you wish to change your behavior, you may need to change your broader environment such as by changing your job, moving to a new location, or becoming more active in changing the system you are in. The same holds for others too. If you wish to influence what they do, you may need to change their broader environment!

Highlight 8.2 Broader Environmental Effects

"What we are is shaped both by the broad systems that govern our lives—wealth and poverty, geography and climate, historical epoch, cultural, political and religious dominance—and by the specific situations we deal with daily. Those forces in turn interact with our basic biology and personality."

(Zimbardo, *The Lucifer Effect*, 298)

Facets of Your Broader Environment

Your Cultural and Historic Context

Culture refers to the distinctive patterns of behavior, values, beliefs, and technologies found in a society or other group, as indicated in Chapter 5 when discussing regular and routine behavior. The major culture of the society around you refers to an enveloping environment that has affected you in the past and influences you now. Some examples of effects are discussed below.[2]

For instance, your parents and other caretakers learned from others to speak the language that they used with you. They learned certain "normal" behaviors and values that then affected what you learned and did—either by teaching you or through other socialization techniques ranging from smiles and compliments to rejection and possible ostracism. In other words, those around you helped establish many of the references you now have concerning what you and others do. They also influenced your knowledge, skills, and attitudes. But what they did was influenced by their own upbringing and experiences.

Such broad environmental effects, indirect though they may be, operate in other ways too. For example your status in society is largely dependent on the dominant values of people in your society. If you were born female, your status will be different than if you were born male—being of lower status and having less power in many cultures. If you were born to royalty, in some societies you have high status and more power. The same holds in some societies—but not in others—if your parents were wealthy or if you are a well-educated professor or medical doctor. For example, during the cultural revolution in China, you would have been despised by many people if you were a rich landholder, or from a wealthy family, or a "decadent" professor.

Highlight 8.3 Cultural Homogeneity

Cultures and subcultures are expected to be "relatively homogeneous internally in the following respects: the types of settings they contain, the kinds of settings that persons enter at successive stages of their lives, the content and organization of... activities, roles, and relations found within each type of setting, and the extent and nature of connections existing between settings entered into or affecting the life of the developing person. In addition, these consistent patterns of organization and behavior find support in the values generally held by members...."

(Bronfenbrenner, *Ecology of Human Development*, 258)

Depending on your culture, your kinship ties may also expose you to the consequences of actions taken by family members or others with whom you associate.[3] If a parent, child, brother, or sister is a murderer, convicted sex offender, or has otherwise failed badly given the standards of your culture, the way you are treated by others may change. Perhaps you will be treated as tainted or foul in some way. Or you may be expected to do certain things, such as to severely discipline or even kill your sister or daughter—as occurs in some traditional Middle East cultures where male relatives are expected to act this way against a relative who has dishonored the family.

Your status and role may also change, depending on the actions of others. For instance, if your daughter has a child, you become a grandparent. If she is unwed, you may be branded as a bad parent. Your daughter, in turn, becomes a mother, with all the rights, privileges, and status accorded to that position by people in her environment. Similarly, in an organization, if a subordinate or team under you, even one that you have never met, does something extremely well or poorly, that action may reflect on you—either to your credit or detriment—to the extent that you may be praised and rewarded or disgraced and forced to transfer or resign.

Religion is another aspect of the broader environment that has affected the values and behavior of many people, including you. The Golden Rule, the Ten Commandments, Hindu rituals called for upon the death of a relative, taboos such as incest, and other norms and rules may have influenced you directly or indirectly. This can happen through their effects on people who have attempted to socially control your behavior either directly, through rewards and punishments related to their norms and rules, or indirectly, by passing laws grounded in religious morality that affect what you do. For example, even if you are not religious, the religious commandment "thou shalt not steal" is embedded in laws about theft—laws that probably affect what you do, especially if you may be caught stealing.

Also, your development and that of others with whom you interact, including your parents, teachers, and bosses, have been embedded in and shaped by the conditions and events occurring during the historic period in which you and they have lived.[4] Such conditions and events as the 9/11 Twin Tower attacks in New York City, recessions, and wars have also affected what you and others do. These and other historic events and conditions are all part of the broader environment—an environment that has affected your development, options, and opportunities for action and that of other people. Here are some more examples:

- If you were born 200 years ago, you probably would not have thought about the right of women to vote. They didn't vote in most places. Today, the right of women to vote is championed around the world.
- If your country's leaders declared war on a major scale, you were probably affected by inconveniences caused by the war effort or perhaps by disruptions in your life caused by service in the military.
- Easy access to information and the ability to communicate quickly and widely via phones, texting, e-mail, and other computer internet systems probably affect things that you do now as compared to what was possible some years ago.
- You are probably more comfortable using some of the more recent technologies available than are people much older than you.

Finally, as discussed in Chapter 5, the social norms of the broad environment surrounding you in various behavioral settings also affect what you do. If you work in an organization, norms concerning suitable dress and attire, the formality of behavior with others, what is considered to be a proper day's work, and the freedom to speak out assertively probably affect what you do. Similarly, the norms of behavior at stores, restaurants, schools, libraries, and highways also define and limit your behavior.

Highlight 8.4	L. S. Vygotsky's Thesis

"...Vygotsky... set forth the thesis that the potential options for individual development are defined and delimited by the possibilities available in a given culture at a given point in its history."

(Bronfenbrenner, *Making Human Beings Human*, 100)

With this as background, let's move on to some other facets of your broader environment that may influence what you do. Although these are interacting parts of your broader environment, we'll look at them separately to help highlight their potential influence and effects.

Your Physical Location

Where you live, work, or go to school also affects what you do by limiting or expanding opportunities available to you. In a rural community, you may be surrounded by a relaxed living style, less crime, have fewer resources available at school, have no medical specialists nearby, and face a lack of public transportation, inadequate childcare, and fewer services compared to living in an urban environment.[5] As a result, many of your options are more restricted than if you live in an urban community. But some possibilities are expanded. For example, rural schools tend to be smaller than urban schools. And so, as discussed in Chapter 6, if you go to a small high school, you may participate more frequently in a greater variety of activities because there are fewer students to play on sports teams, cheerlead, participating in school plays, and do other things.

Location can affect people in other ways too. If you are a citizen of Yemen, Iran, or North Korea, you may be treated differently at airport security screenings in Europe or the United States, than you would be as a U.S. or British citizen.

In short, the location of the broader environment in which you live and have lived makes a difference in what you experience, think, and can do. Some of these differences are due to characteristics of the location. Others are due to stereotypes and prejudices associated with how individuals think about people from that location—and have nothing to do with you as the person that you really are.

Economic Conditions

Economic conditions have affected you in the past and also what you do now. Whether you live in a poverty-stricken or wealthy area, live during a recession or economic boom, and whether a company where you or another family member works is doing well or poorly economically, can influence you and what you do.

For example, if you live in an urban slum, your life may be more hazardous and stressful and you will be treated differently by others than if you live in a wealthier part of town. Besides gang violence, drug trafficking, crime, and a lack of employment opportunities for yourself, if many people around you are unemployed, they will behave differently than if they are working regularly in stable jobs. As a result, role models who are routinely unemployed may not exhibit and help you develop habits associated with steady employment—such as getting up and doing things "on time." If you are raised in a poverty-stricken, slum environment, you are also more likely to consider violence as a way of life.[6]

Highlight 8.5 **The Truly Disadvantaged**

"The concentration of poor people in poor areas increases all kinds of stress, deprivation and difficulty—from increased commuting times for those who have to leave deprived communities to find work elsewhere, to increased risk of traffic accidents, worse schools, poor levels of services, exposure to gang violence, pollution and so on."

(Wilkinson and Pickett, *The Spirit Level,* 163)

Even if your environment is not poverty-stricken, recessions can affect you and may have affected you profoundly. When you were growing up, a poor economy overall or within an organization may have affected your parents. Perhaps your father or mother lost a job and couldn't find work. This may have led to stress and financial problems within your family. It may also have resulted in greater domestic roles or outside employment for you as a way of compensating for less money coming into your family. It may have restricted your ability to obtain higher education and your later ability to marry well. Your father may have left your household for extended periods of time, looking for and possibly finding work in distant cities. Perhaps your family had to move elsewhere in order to work. In addition, if your parent or parents were away working most of the time in order to earn enough money to make ends meet, you may have spent more time with your peers for better or for worse.[7]

Even today, if you work for money, poor economic conditions may cause you to lose your job or make your life more stressful by having you think about possibly losing the job that you have. If you are young, poor economic conditions may make options such as military service seem more attractive. If you are employed, such conditions may also affect whether your organization is able and willing to provide you with job training that you may need to work better or to advance within the organization, since during poor times, training budgets are usually among the first items to be cut within an organization. If labor elsewhere is cheaper or other costs of doing business less expensive, you may even lose your job as businesses relocate their operations to less expensive locations.

In short, even if you are not affected directly, economic conditions can affect the way you are treated by other people, the opportunities available to you, and your levels of stress and happiness. The same holds for everyone else too as Highlight 8.6 indicates.

Highlight 8.6 — The Economy and Behavior

"As the economy falters, the incidence of admissions to mental hospitals and prisons, child abuse, spouse abuse, suicide, and so forth are dramatically increased . . . ."
(Dziegielewski and Wodarsk, "Macro-Level Variables," 261)

Organizations

Almost every stage of your life has occurred within an organizational context and still does.[8] You may have been born in a hospital; married in a church, temple, or mosque; been a Boy Scout or Girl Scout; received education at a school; been employed by a business or government; bought food, clothing, and other items in stores; perhaps served in the military; or had contact with other public and private organizations. If so, aspects of those organizations, such as their goals, policies, and procedures undoubtedly affected what you did, could do, and what happened to you. The following indicates some ways this occurs.

Governments

The governments around you, both local and national, affect you in many ways. Most of this influence is a result of actions taken by people who don't even know that you exist or who may have died many years ago. Here are a few examples:

The Law and Related Policies: Laws are made by governments to guide, control, and sometimes help people.[9] Those laws and related public policy are part of the broader environment that shapes the immediate environment surrounding you and others during your everyday life.[10] Nearly every law and policy of government has either a direct or indirect effect on families and on you.[11] For instance, if you drive a car, whether you wear seat belts or not, the speed at which you drive, the side of the road that you drive on, the alcohol levels of your blood when driving, and the taxes that you pay when buying your car all influence what you do, since if you violate laws concerning these matters, you are subject to fines, arrest, or other punishment by the government. The same holds for other people too.

In many ways your quality of life is also affected by laws and government policies. The food that you eat should meet certain government standards. The income that you obtain and the work that you do is affected by labor laws such as minimum wage and safety standards. Medical care that you obtain is similarly affected, for example, by the licensing of medical personnel and approval of certain drugs. If you are poor or elderly, expenses of your health care may be covered by programs such as Medicare in the United States or national health services in most other developed countries. Similarly, retirement benefits may be provided by Social Security or other retirement plans regulated by government laws and rules.

Here are some other examples that indicate the variety and extent of government lawful influence:

- Government enforcement mechanisms, including police officers, inspectors, courts, and prisons, help ensure that you obey laws that have been established.
- Government policies concerning traffic safety, including safety features on cars, restrictions on teenage driving, and driving under the influence of alcohol, have reduced traffic death rates.[12] Similarly, government bans on smoking in public places have had positive effects in reducing levels of smoking in smokers.[13]
- Higher taxes on alcohol help reduce binge drinking.[14]
- As local, state, and national governments have extended gambling venues and options such as lotteries in search of new revenues, the numbers of pathological gamblers have increased proportionately.[15]

- A California statewide policy bans certain forms of athletic training during unhealthful air conditions. Urban planning in some places also promotes healthful activities such as walking and bicycling to school by building suitable pathways.[16] Warning labels on cigarette packs and public service advertisements about the hazards of smoking are other attempts by governments to influence peoples' behavior and health.
- Maternity and paternity leave in many countries is also controlled by laws. For example, over 125 countries provide job-protected, child-birth-related leave, averaging 16 weeks, including pre-birth and post-birth time off. In the United States, a parent can take up to 12 weeks of unpaid, job-protected leave each year. In Sweden, parents are entitled to 390 days of paid leave that they may use up to the child's eighth birthday—leave that they may use as they wish, either monthly, weekly, daily, or even hourly.[17] In addition to newborn leave, many countries also provide child-care services at little or no cost for working parents of infants and children.[18]

During 2003, a scandal about the torture of Iraqi insurgents by U.S. soldiers in Iraq was unveiled. A few years later, Philip Zimbardo in his book, *The Lucifer Effect*, charged President George W. Bush, Vice President Cheney, Secretary of Defense Rumsfeld, and down through the hierarchy of command in the Bush Administration, with laying the foundation for the abuses that occurred. Zimbardo wrote that Bush and other government officials did this, in part, by redefining torture as an acceptable tactic in "the war on terror." Other actions included exempting captured insurgents under military arrest from safeguards provided by the Geneva Conventions, issuing and endorsing related policies, and ignoring cases of abuse in Iraq, Afghanistan, and Cuba.[19]

Similarly, the U.S. mortgage default crisis and associated Great Recession of 2007 to 2009, "could not have occurred without government-led deregulation, expansionary monetary policies, and goals to expand mortgage financing to marginal borrowers," according to John Liederbach, an Associate Professor in the Criminal Justice Program at Bowling Green State University in Ohio.[20]

Highlight 8.7 "The System" and Behavior

"The System consists of the agents and agencies whose ideology, values, and power create situations and dictate the roles and expectations for approved behaviors of actors within its spheres of influence . . . ."

"'Bad systems' create 'bad situations' create 'bad apples' create 'bad behaviors,' even in good people."

(Zimbardo, *The Lucifer Effect*, 445, 446)

The Constructed Environment Around You: Part of your environment has been built by local, state, and national governments. The roads, street signs, traffic lights, subway and bus systems, and the money you use exist because of government decisions. These structures and systems influence your life both directly and indirectly. Directly, for instance, since you drive on those roads or use public transportation to go from one place to another. Indirectly, for example, by permitting truckers and delivery men to easily move goods from far away to stores near your home—goods that you then purchase and consume.

Governments also construct buildings such as schools, public libraries, post offices, and facilities at parks. Since you have used some of these and will continue to do so, your life would be different if these facilities did not exist.

Decisions Made by People in Governments: Besides decisions to establish laws, construct public buildings, and establish transportation systems, many other decisions made by those in government affect you. The results of these decisions can be subtle and unexpected. For example, the decisions to go to war in Iraq and Afghanistan resulted in a dramatic increase in people diagnosed with post-traumatic stress disorder (PTSD) within the United States.[21] Wars, of course, also affect people in many other ways, including the news to which you are exposed and possible dislocations caused by family members and friends who are affected.

Similarly, broad economic decisions such as the North American Free Trade Agreement (NAFTA) have had an impact on towns and communities when companies in them left the United States to open plants in Mexico.[22] Subtle impacts that may be less obvious include lower prices on goods and services, job creation due to the greater export of American goods, and greater investment in the United States by Canada and Mexico that has contributed to tax revenues and jobs.[23]

Even the demand for war reparations presented by the governments of France and Great Britain to Germany in 1921 had serious effects. The bill for about US $33 billion caused terrible inflation in Germany, where the mark, which had been four to the dollar, fell to 4,000,000,000 to the dollar by 1923. As a result, Germans lost their savings, groceries cost billions, hunger riots broke out, and the Nazi movement began to take action that eventually affected much of the world.[24] Such effects continue to influence you and other people.

Schools and Education: The school system is another important part of the environment affected by government decisions and laws. Other people set the curriculum for students to learn what they consider to be important. In addition, everything from teacher qualifications, school hours, costs of schooling including the taxes you pay to support the school system, the ages of children who may attend, and even the hours and number of days of school are based on government decisions, rules, regulations, and laws.

Larger Businesses, Corporations, and Other Employers

Employers influence you and others in many ways—in fact, they must do so if their organizations are to continue existing. For example, corporations and other large businesses have goals to be achieved, policies to guide behavior, and procedure manuals to be followed by employees. Job descriptions affect what workers do and are expected to do. Employee evaluation procedures provide feedback to employees to help ensure that they are doing what the employer expects.

Work environments are often designed to help workers better achieve an organization's goals. The architecture may even be arranged to produce positive feelings and reduce stress, such as by imitating natural settings and elements.[25] In addition, salaries, bonuses, and incentives are provided to influence worker behavior.

The operations of organizations are in turn affected by their broader environments. For example, the overall economy, changes in customers and clients served, and changes in technology can all affect an organization. So can the availability of manpower for recruitment, new legislation, political change, the activities of special interest groups, and the plans of competitors and other organizations. These influences can then, in turn, affect you and others with whom you interact. For example:

- Poor economic times for an organization can mean budget cuts and a loss of your job or of that of a spouse or parent.
- A lack of skilled persons who can be recruited may mean that an organization's present employees will have to be trained and assigned to those skilled positions.[26]
- The emergence of AIDS and other diseases have affected the training needs of personnel working in health units.

Such external factors, besides ultimately affecting what employees like you may do, also affect others, such as the parents of children, resulting in indirect effects on the children themselves. This may have happened to you for example when a parent could not spend time with you because he or she had to work late, or came home stressed after a hard day at work.

Similarly, even if you are not employed by a business or corporation, you may be affected by those businesses. For example, profit-making news organizations often sensationalize crime, with the result that they are major producers of fear in the American people.[27] This occurs, in part, since sensationalism attracts attention and is good for the news business, which can then make greater profits through increased sales to advertisers and subscribers. Unfortunately, however, what is emphasized and reported by news agencies affects people's perceptions of reality. Events unduly emphasized stimulate misperceptions as does a lack of news and information—misperceptions such as the belief held by many Americans that "the U.S.A. has the best health care system in the world." This presumed fact has been clearly contradicted by a report showing that the United States ranks quite poorly when compared to 16 other high-income countries.[28]

Other Organizations Including Nonprofit Groups

Many of the ideas just discussed also apply to other kinds of organizations. For example, the behavior of people within nonprofits is guided by policies, job descriptions, plans, and procedures for accomplishing goals. These influence what people within them do. Sometimes their clients are also affected by these factors—such as the people helped by food kitchens, homeless shelters, the Salvation Army, and other charities.

Some of these organizations have even changed behavior on national and international scales. For example, Mothers Against Drunk Driving (MADD) in the United States and Canada has helped reduce drunk driving by increasing awareness of the problem, by influencing passage of drunk driving laws and laws concerning minimum drinking ages, and by helping victims of drunk drivers. Similarly, the influence of nonprofit organizations such as the Red Cross and Red Crescent, the Boy Scouts and Girl Scouts, Amnesty International, UNESCO, Rotary International, and Habitat for Humanity have been great.

All of these organizations form part of your broader environment and may have affected you, too.

Technology and Science

Another dimension of your broader environment is science and technology. Compared to earlier times, you do many things as a result of the science and technology that now exist.

For example, you may now quickly drive rather than walk to go someplace—something that was not possible much more than a hundred years ago except by horse or train. You may even fly from place to place. Likewise, you may have a smart phone, telephone, or other communication device that you use. The same for a computer that you may use to gather information, contact friends, work, or play games. Perhaps you have a wristwatch, clock, or other device for telling time. You may look at television or listen to the radio for news and entertainment. You turn on lights to see better at night. Maybe you now read digital books, magazines, and newspapers, instead of looking at paper versions.

How about the birth control pill? That has had a great effect on the behavior of many people, perhaps on you too. So have other inventions and technological developments—the atomic bomb, for example. If you lived some years ago, perhaps you or your parents stored an emergency supply of food at your home or even built a bomb shelter in case there was a nuclear war.

In addition, the way that many devices are designed forces you to behave in certain ways if you wish to use them. You may have to turn a machine on first, press certain buttons or adjust dials in a particular way, and do things in certain sequences in order to use many of these tools, appliances, gadgets, and technical devices.

Technological developments have also opened up new possibilities for action. Instead of going to someone's house or office to talk with them, you can now call them on the telephone or use your computer to do so. Instead of writing a letter, you can send an e-mail or text message. Instead of asking for directions when driving to a new place, you can use a GPS, smart phone, or software such as MapQuest on your computer to find your way. More broadly, television, radio, internet news, and related advertising have transformed political campaigns beyond recognition compared to the times before those technologies existed.[29]

In short, the technology around you and the scientific advances of our time have greatly influenced the things that you, I, and other people do and can do.

Other Broad Environmental Influences

We can also distinguish other broad environmental influences on your life and on the lives of others. For example, some that overlap with those mentioned before, but are worth discussing separately given their importance, include the mass media, the medical system, and natural occurrences.

The Mass Media

The mass media and those behind them, including business managers and politicians, have an enormous influence on people's knowledge, thinking, and behavior. Television, radio, newspapers, magazines, movies, billboards, and the internet are looked at or listened to by many people every day. These media affect a great deal of what you and others know and believe about the world. They are, for example, the reason why many people are fearful of crime and do not shop, work, or walk around some cities late at night. They are a reason why you and others buy various advertised products. And the media are a reason why many people are afraid of terrorism, child abduction, certain ethnic groups, and even extremely rare possibilities such as razor blades or poison being put in Halloween candy given to children in the United States. As indicated before, mass media have also transformed the way that politics and political campaigns are conducted.[30]

Although a great deal of what we know about the world has been obtained through television and other mass media, the content is often not a true reflection of the real world. What is presented is selected and interpreted by those producing what we see, hear, or read—those who, in the case of commercial media, need to make money through advertising in order to survive. To make money, large audiences are required. The larger the audience, the greater the advertising and potential profits for producers. As a result, what is presented on commercial television is selected to attract and satisfy viewers and advertisers, rather than to faithfully reflect reality.

For example, factors that TV producers use to select the news that they present, include:

- **Personalization**: A newsworthy story is often about individuals, such as the head of state, a movie star, a serial killer, or a terrorist.
- **Drama and Conflict**: Conflict-filled, violent, and dramatic events are often overreported.
- **Action**: Newsworthy events often contain action and observable occurrences.
- **Novelty and Deviance**: Events outside the predictable range of news have a better chance of being covered.
- **Inoffensiveness**: The story must not be too offensive to what people wish to hear. They also should not offend advertisers or the parent company of the news channel.
- **Credibility**: The story presented must be viewed as credible.
- **Sound Bites**: The story must be packable in small pieces, fit for brief TV news.[31]

The result is that what we know about the world, especially that with which we do not have direct contact, is largely shaped by the media to which we are exposed—the contents of which are selected by unknown people for reasons unknown to us. For example, what do you know about China? Probably, for the most part, what you have seen, heard, or read on different mass media. And much of this is undoubtedly distorted. Such distortion struck me when President Richard Nixon visited China in 1972. The images broadcast on TV were nothing like those that I had expected, based on previous media exposure, slanted reporting, and Cold War propaganda.

Of course, the mass media can also be used to promote desirable behaviors, such as those concerning better health (for example, "safe-sex" and good eating practices), improved safety (such as the use of seat belts), and vocabulary growth, pre-reading skills, and nonracist attitudes by children (via "Sesame Street").

Highlight 8.8 The Mass Media and Behavior

"People read newspapers. Despite everything we think we know, their contents seep in, we believe them to be true, and we act upon them, which makes it all the more tragic that their contents are so routinely flawed."

"Dr. Benjamin Spock ... recommended that babies should sleep on their tummies. ... we now know that this advice is wrong, and ... has led to thousands, and perhaps even tens of thousands, of avoidable cot deaths. The more people are listening to you, the greater the effects of a small error can be."

"Studies looking at the use of specific health interventions before and after media coverage ... found that favourable publicity was associated with greater use, and unfavourable coverage with lower use."

(Goldacre, *Bad Science*, 108, 324)

"Q: How many murders has the average [American] child seen on TV by the time he or she finishes elementary school? A: Eight thousand, plus 100,000 other acts of violence ...."

"The evidence overall strongly converges on the conclusion that media violence does have harmful effects, especially on children, primarily in three areas. After exposure to media violence, there is an increase in fear, violent behavior, and desensitization ...."

"Even if media violence is responsible for only 5 to 15% of societal violence ... that is still very important. Because ... even a very small effect of media can be substantial in terms of numbers."

(R. Harris, *Cognitive Psychology*, 257, 259, 288)

The Medical System

The medical system deals with matters of health and wellness, including the attempted social control of people who deviate from accepted norms of behavior. Problems such as gambling and alcoholism that were once considered moral problems are now considered medical problems, at least in many Western societies. Such deviants tend to be viewed as sick and in need of treatment or rehabilitation. This kind of medicalization of deviant behavior is another broad environmental influence on what people do. Such influence includes the effects of mental hospitals, psychiatry, drugs for depression and anxiety, self-esteem training, and public health approaches such as campaigns to remove or reduce environmental hazards like lead paint from homes, businesses, and schools.[32]

Medical rules including those concerning vaccinations required to attend school or travel to other countries, and physical examinations needed to work, immigrate, or obtain life insurance, may also affect what you do or have done in the past. Norms of behavior linked to visits to doctors' offices or stays in hospitals may have also.

Highlight 8.9 **Some Medical Meanings**

Medicalization "is the process by which personal and social problems are redefined as psychiatric or medical problems."

(Chriss, *Social Control*, 64)

Medical social control refers to "the ways in which medicine functions (wittingly or unwittingly) to secure adherence to social norms—specifically, by using medical means to minimize, eliminate, or normalize deviant behavior."

(Conrad and Schneider, *Deviance and Medicalization*, 242)

Natural Occurrences: Mother Nature

Many natural occurrences far removed from us affect what we do. For example, a storm knocks out power to a day care center, causing the center to be closed for the day, forcing the parent of a child to stay home rather than go to work. Similarly, the presence or threat of a disease to a community, droughts, the occurrence of a rainy season such as the South Asian monsoon, and dramatic events such as a distant hurricane, earthquake, famine, or volcano eruption can all affect what you do or are able to do. For example, a volcanic eruption in another part of the world may result in canceled airplane flights due to ash in the air. Or the ash may result in beautiful sunsets that may inspire you to linger and watch it, perhaps share it with a lover, or take a picture or make a painting of it. An eruption can even cause temporary global cooling—as occurred after the explosion of Krakatau in Indonesia during 1883—perhaps forcing you to shovel more snow during a harsher winter than usual or eat fewer citrus fruits such as oranges due to damage done as a result of the colder weather.

In short, many things both near and far affect what you experience, do, and can do.

Preview of the Next Chapter

Based on this and the previous chapters, you should now have a broad idea of how both you and your environment influence your behavior. In the next chapter, we'll take a closer look at some of the theories and models of behavior that have been developed to help explain human behavior. You'll finish with an improved understanding of the thinking emphasized in this book and why some explanations are better than others.

Further Reading

For a classic introduction to the immediate and broader environment, look at:

- Urie Bronfenbrenner, *The Ecology of Human Development: Experiments by Nature and Design* (Cambridge, MA: Harvard University Press, 1979).

Social workers are more sensitive than most professionals to environmental effects on behavior and related interactions. An informative text in this regard is:

- Rudolph Alexander, Jr., *Human Behavior in the Social Environment: A Macro, National, and International Perspective* (Los Angeles: Sage, 2010).

To better understand the influence of context on behavior and why good people do bad things, see:

- Philip Zimbardo, *The Lucifer Effect: Understanding How Good People Turn Evil* (New York: Random House: 2007).

For information about U.S. government policies concerning children and families, go online to:

- The Clearinghouse on International Development in Child, Youth and Family Policies, Columbia University, at http://www.childpolicyintl.org.

A book often cited to show effects of the U.S. economy on behavior is:

- Glen H. Elder, Jr., *Children of the Great Depression: Social Change in Life Experience* (Boulder, Colorado: Westview Press, 1999). First published 1974 by the University of Chicago Press.

Examples of how decisions made by government leaders can affect people are contained in:

- Ian Kershaw, *Fateful Choices: Ten Decisions That Changed the World, 1940-1941* (New York: Penguin Press, 2007).

For information about the effects of mass media on the knowledge and behavior of broad audiences, see:

- Richard Jackson Harris and Fred W. Sanborn, *A Cognitive Psychology of Mass Communication*, 6th ed. (New York: Routledge, 2014).

For insights about how politicians, the media, and activists manipulate and misinform you and others, see:

- Wayne Le Cheminant and John M. Parrish, eds., *Manipulating Democracy: Democratic Theory, Political Psychology, and Mass Media* (New York: Routledge, 2011).

Endnotes

1. Technical terms that have meanings similar to "the broader environment" as used here include the macro-environment, the macro-system, and the exosystem.

2. Culture refers to aspects of the social and physical environment. Those aspects are what affect behavior, not the abstract word "culture" that refers in a general way to them. This distinction between an abstract word and what the word refers to is clarified in the next chapter. Unfortunately, many social scientists do not seem to be aware of the difference and treat abstract words, such as culture, as causal entities that directly influence behavior.

3. Elder, "The Life Course Paradigm," 118.

4. Bronfenbrenner, "Developmental Ecology," 641-643.

5. Alexander, *Human Behavior*, 283-284; Kirst-Ashman, *Human Behavior, Communities, Organizations*, 244-245.

6. Moen, Elder, and Lüscher, *Examining Lives*, 532-534.

7. Elder, *Children of the Great Depression*; Bronfenbrenner, *Ecology of Human Development*, 269-284.

8. Alexander, *Human Behavior*, 183.

9. Those who make and pass laws do so to satisfy goals and other references that they have—as indicated by Perceptual Control Theory.

10. Bronfenbrenner, *Human Development*, 9.

11. Bogenschneider, *Family Policy Matters*, 80.

12. Traynor, "Impact of State Level Behavioral Regulation, 425."

13. Orbell and others, "Social-Cognitive Beliefs, 753."

14. Boston University Medical Center, "Alcohol Taxes."

15. Nathan and Gorman, *Guide to Treatments,* viii.

16. Gorman et al., "Designer Schools, 2522."

17. Bennhold, "In Sweden"; also see http://www.childpolicyintl.org, the website of The Clearinghouse on International Development in Child, Youth, and Family Policies at Columbia University.

18. Table 1.22, Section 1.2: Early Childhood Education and Care, The Clearinghouse on International Developments in Child, Youth, and Family Policies, Columbia University, accessed July 28, 2010.

19. Zimbardo, *The Lucifer Effect,* 378, 380-443.

20. Liederbach, "'Pass the Trash," 34-35.

21. Nathan and Gorman, *Treatments,* viii.

22. Alexander, *Human Behavior,* 83.

23. *NAFTA at Twenty.*

24. "The Rise of Adolf Hitler," The History Place, accessed December 12, 2010, from http://www.historyplace.com/worldwar2/rise ofhitler/putsch.htm

25. Joye, "Architectural Lessons, 325."

26. Pfau, *How to Identify Training Needs.*

27. Alexander, *Human Behavior,* 58.

28. National Research Council and Institute of Medicine, *U.S. Health in International Perspective.*

29. Harris, *Cognitive Psychology,* 235.

30. Ibid., 225-256.

31. Ibid., 191-196.

32. Chriss, *Social Control,* 64-86.

PART IV

Behavior Theories, Analysis, and Change

This part consists of three chapters. The first explains why this book emphasizes certain views and theories of behavior. The next presents a slightly expanded view of Perceptual Control Theory that summarizes much of our discussion so far. Guidelines are also provided that will help you systematically analyze your own behavior. The concluding chapter builds upon what has been presented before and explains how to change your behavior if you really want to do so.

CHAPTER 9

THEORIES AND MODELS OF HUMAN BEHAVIOR

The Big Picture

Hundreds if not thousands of theories and models of human behavior exist. Some simply summarize observations made and regularities noticed about certain kinds of behavior. Some are aimed at explaining why certain behaviors occur. Many focus on the individual person and ignore the environment around that person. Others focus on the environment and ignore what happens inside the person, treating him or her as an unspecified "black box" having inputs and outputs, stimuli and responses. A few consider the person and environment as an interacting system or unit. All are abstractions of what actually exists or occurs, although some seem to be closer to describing reality than others.

Most concepts used in these theories and models are abstractions that have no existence in physical reality, although they are presumed to have value in describing or explaining what people do. A major problem is that many of these abstractions are treated as "real things" that cause people to do what they do. As a result, abstractions that do not really exist as entities such as attitudes, interests, the ego, and culture are treated by many social scientists as real things that cause people to act in certain ways. As a result, we see or hear statements such as "attitudes cause behavior" and "culture causes behavior"—statements that at a superficial level seem to explain why people do things, but actually explain nothing in causal terms. Although such statements may be OK for everyday "folk conversation," they are often misleading and usually cannot be verified or put to practical use very effectively.

Another problem with many theories and models is that they treat behavior in an outdated, simplistic way. They treat behavior as if it is caused in a one-directional, linear manner—in its simplest form, for example, as if something called "a stimulus" or other "input" causes you and others to "respond," behave, or have "output" in a certain way. Such linear thinking is usually much too simple, since you don't just react to your environment as if you have no control over what you do. You are much too complex for that. You *interact* with your environment and some of your environment *interacts* with you. That is, you are constantly adjusting to and being influenced by what is around you, and what is around you is often influenced by you.

As previously discussed, what you do is influenced by (a) your internal organization and structure including your physical abilities, (b) the environment around you, and (c) the continuous interaction you have with your environment. These "facts," if we may call them that, make environmentally interactive theories and models look extremely attractive, which is why such views are highlighted and focused upon in this book.

Theories and Models of Behavior

An Overview of What You Have Learned So Far

Based on the preceding chapters, you have seen that many factors influence your behavior. Some of these factors are internal to you, such as your perceptions, error signals, hormones, and skills. Some are external, in your environment, such as the people around you, physical objects, the weather, and government laws and policies. Some are the result of interactions between you and your environment, such as your health and what you know and have learned about the world. Highlight 9.1 summarizes these influences.

Highlight 9.1 indicates the many internal and external influences upon you, your perceptions, and behavior. At the center of these influences is your body, represented in more detail by the lower part of the diagram. As discussed in earlier chapters, what you do depends on the way that you are structured and organized, as illustrated at the sub-system level by Powers' Perceptual Control Theory (PCT) that focuses on your nervous system. Also, your body's structure and organization as well as your behavior are influenced by your environment, as indicated by the upper part of Highlight 9.1. (For a more detailed discussion and description of the different hierarchical levels shown, see Appendix 1).

Highlight 9.1 Hierarchies of Influence and Understanding

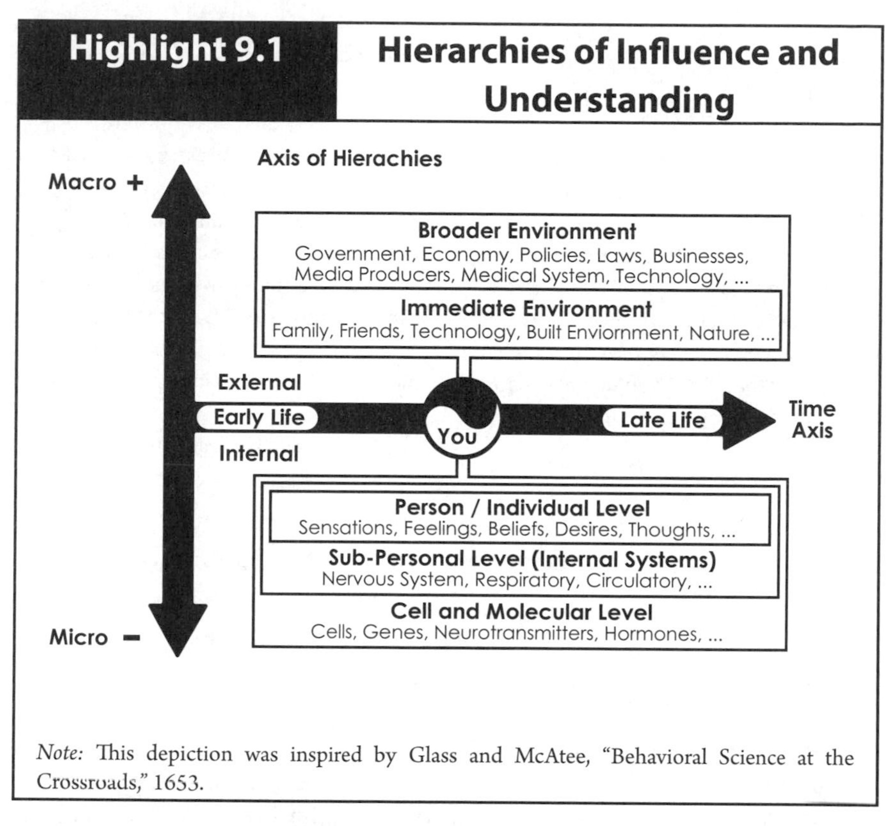

Note: This depiction was inspired by Glass and McAtee, "Behavioral Science at the Crossroads," 1653.

However, nearly all that you do is based on "negative feedback control." This concept, summarized in Highlight 9.2, is the basis of Powers' PCT model. *Negative feedback* refers to the fact that your body is organized so that feedback concerning your actions helps you reduce error signals that occur—error signals that indicate a difference between what you perceive and your references. That feedback is needed to achieve or maintain the references that you have, such as your goals and essential physiological states, because the feedback indicates how well you are doing. That is, feedback helps you determine whether you are increasing or decreasing error signals that are occurring. You need to know this in order to better direct your activities.[1] Without feedback, you would act blindly. You would not know whether or not what you are doing is working to achieve results that you desire or need to survive.

Highlight 9.2 Negative Feedback Control and PCT

"Negative feedback control as used in PCT can be summed up very quickly. It involves continuously *perceiving* the current state of whatever is to be controlled, continuously *comparing* that perception with the intended state, and continuously *acting* to reduce the difference between perceived and intended states, keeping the difference as close to zero as possible. The feedback part of negative feedback says that effects of the output (action) caused by the difference feed back to the input to change the perceptions and affect the difference. The negative part comes from the fact that the action decreases the difference instead of increasing it. Causation runs around a circle or loop."

(Powers, *Living Control Systems III*, 7)

We'll come back to Powers' model and theory a bit later. But to really appreciate it and what you do, let's look at a few more concepts. Specifically, we'll consider systems theory, circular loop causation, abstraction and reality, and the overgeneralization of psychological research that often occurs.

Systems and Holistic Views of Behavior

A system is a set of interrelated parts. Systems range in size from small, like a cell, to increasingly larger, such as a nervous system, your body, a family system, an organization, a nation, and the world. As these examples indicate, larger systems are made of smaller systems. Also, an important characteristic of a system, by definition, is that the parts of a system are related. They affect one another.

When it comes to your behavior, you and your environment are of special importance. That is, you, your environment, and the relations between you and your environment are special because, together, you form a system—an "organism-environment system."

This system is not always the same. It changes over time. For example, internally, your body changes. From conception to birth to your eventual death, your brain and body develop, many of your cells die and are replaced, your blood glucose levels go up and down during the day, and other changes occur. Externally, your environment changes, too—from you being surrounded as a fetus by your mother's womb; and then as a child to your parent's home and care; to being surrounded by schools, workplaces, stores and other behavior settings; to daily weather, new laws, recessions, and wars. And the relationships between you and your environment change from moment to moment, daily, and yearly, as do relationships between different parts of your environment, and between parts of your body.

The complexity of all of your body's different systems and of the broader environmental systems of which you are a part is impressive—even overwhelming. So many things, events, and relationships affect so many other things, events, and relationships. To get a sense of this, see the analysis of factors affecting obesity and related eating behaviors shown in Highlight 9.3. As you may begin to realize by looking at the many factors and interrelationships between them, the system that affects how fat you and others are is extremely complex. Each of the following environmental factors is a potential influence:

- Education, such as programs to increase food literacy and food skills; and healthy school meals
- The Media, including advertising to children, and health warnings on TV
- Technology, such as TV and computers that, for example, increase the desirability of indoor leisure activities
- The Nature of Work, often requiring less physical activity now-a-days
- The Built Environment, Recreation, and Transport, such as promotion of walking and cycling
- Healthcare and Treatment Options, such as the availability of treatments and anti-obesity drugs
- Early Life Experiences, including nutrition after birth and parent obesity
- Food Production and Supply, including the availability of healthy and fast food
- Macro-Economic Drivers, such as the price of food, advertising, and income distribution.

In addition, within you and others there are core biological loops as well as learned control loops that regulate body weight.

Highlight 9.3 The "Full" Obesity System Map

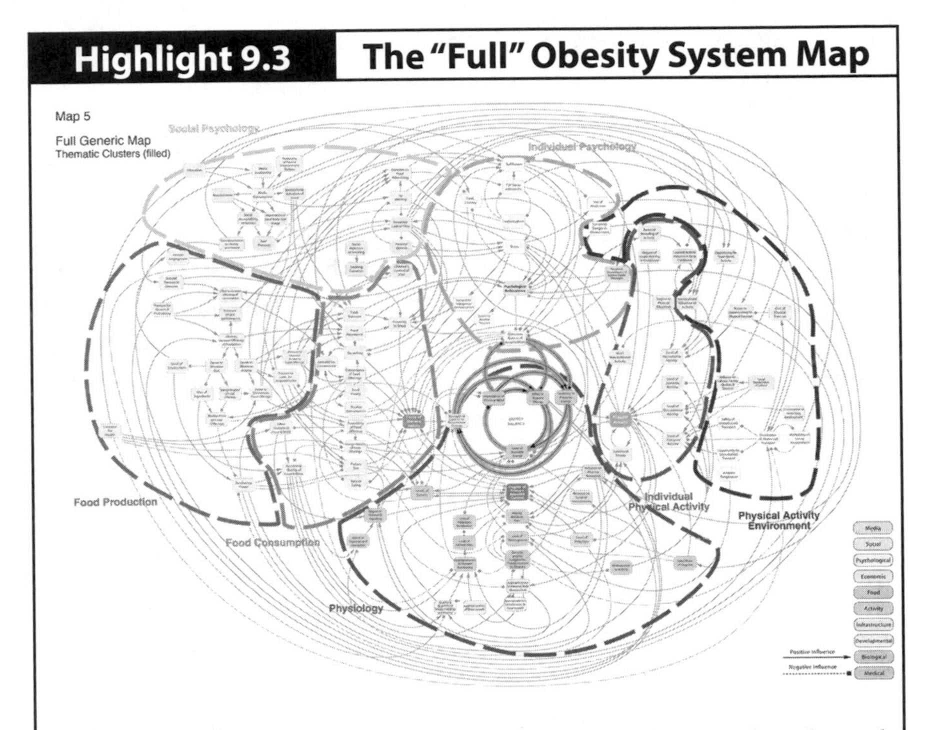

Details of this complex system can be seen online or downloaded from www.foresight.gov.uk and searching for "reducing obesity: obesity system map".

Source: Foresight: Tackling Obesities: Future Choices—Obesity System Atlas, U.K. Government Office for Science. https://www.gov.uk/government/uploads/system/uploads/attachment_data/file/295153/07-1177-obesity-system-atlas.pdf.

The main point is that we need to be aware that we live in a complex world, and when considering why we do things, we should not ignore that complexity. We shouldn't simplify too much. For example, as many people and psychological theories do, we shouldn't focus only on the individual and ignore the surrounding environment that may affect that person.[2] Alternatively, as some practitioners, theorists, and theories do, we shouldn't focus only on the environment and ignore what is happening within an individual and treat him or her like a black box described earlier. We need to consider individuals, their environments, and the interrelationships that occur between individuals and their environments. That is, we need to consider the organism-environment system when trying to understand your behavior and the behavior of others.

In addition, when thinking of systems, it is wise to realize that the characteristics of complex wholes are not reducible to the characteristics of their parts.[3] Just as the properties of hydrogen atoms (H) and oxygen atoms (O) are different from those of water (H_2O), your behavior cannot simply be reduced to the characteristics of genes, nerves, and other parts of your internal systems or to the environmental systems of which you are a part. What you do emerges from the interaction of your various parts between themselves and with parts of your environment, not from the parts themselves. *What you do emerges from how you function as an organism-environment system!*

In short, to understand your behavior and the behavior of others, you need to take a holistic view of yourself, your environment, and how you interact with your environment. You need to consider yourself and your environment as a system whose parts interact and affect one another. The same holds for understanding other people, too. To understand their behavior, you need to consider them, their environments, and the relationships and interactions between them and their environments. See Highlight 9.4 for more details.

Highlight 9.4 Systems and Behavior

"A system is a set of things—people, cells, molecules, or whatever—interconnected in such a way that they produce their own pattern of behavior over time. The system may be buffeted, constricted, triggered, or driven by outside forces. But the system's response to these forces is characteristic of itself. . . ."

"The system, to a large extent, causes its own behavior! An outside event may unleash that behavior, but the same outside event applied to a different system is likely to produce a different result."

"The behavior of a system cannot be known just by knowing the elements of which the system is made."

"There is no single, legitimate boundary to draw around a system."

(Meadows, *Thinking in Systems*, 2, 7, 97)

"The systems scientists . . . .discern relationships and situations, not atomistic facts and events."

(Laszlo, *Systems View of the World*, 9)

Circular Loop Versus Linear Causation

Thinking of the present state of psychology and other social sciences, it is clear that much thinking is stuck in the past and uses an input-output model of behavior that is too simple, out-of-date, and based on an illusion of what is actually occurring. Since the 1930s and 1940s a new model has emerged that describes processes basic to the functioning and behavior of all living organisms—a circular model that falls within the technical area of cybernetics. Let's look at this newer way of thinking as well as the older more linear way. See Highlight 9.5 for some relevant definitions.

Highlight 9.5 Concepts Concerning Cybernetics and Causation

"***Lineal*** describes a relation among a series of causes or arguments such that the sequence does not come back to the starting point."

(Bateson, *Mind and Nature*, 212)

"**cybernetics** *n.* the scientific study of communication and control as applied to machines and living organisms. It includes the study of self-regulation mechanisms, as in thermostats or feedback circuits in the nervous system, as well as transmission and self-correction of information in both computers and human communications....[first defined in 1948 by U.S. mathematician Norbert Weiner (1894-1964)] –**cybernetic** *adj.*"

"**linear causation** the simplest type of causal relationship between events, usually involving a single cause that produces a single effect or a straightforward CAUSAL CHAIN."

"**multiple causation** the view that events, including behaviors, seldom result from single causes but instead from multiple causes working in complex combinations. Multiple causation contrasts with SIMPLE CAUSATION and, often, with LINEAR CAUSATION."

"**circular causality** ... a sequence of causes and effects that leads back to the original cause and either alters or confirms it, thus producing a new sequence, as in a FEEDBACK loop."

"**feedback loop** in CYBERNETIC THEORY, a self-regulatory model that determines whether the current operation of a system is acceptable and, if not, attempts to make the necessary changes."

(VandenBos, *APA Dictionary*, 2nd ed.)

Linear causation is a common view, basic to the physical sciences: X causes Y. Even multiple causation often follows the same kind of one-way lineal logic, with A, B, C, D... and X causing an end result Y. Such thinking is logical. However, when applied to biological systems, it is inappropriate because in living systems, circular, mutual causation occurs. For example, X (your perception) affects Y (your behavior), which affects your environment as well as X (your perception), which then affects Y (your behavior), and the cycle may continue. The reverse effect, from Y to X results from feedback that occurs from your behavior, Y, to your perception, X. For example, when driving a car you may perceive that you are drifting out of your lane (X) and so you turn the steering wheel slightly (Y) and see if you are properly in the lane (X) and if not turn the wheel a little (Y) to adjust your position, see how you are doing (X), hold the wheel steady (Y) if you are properly in your lane, and continue this interactive process as you drive down the road.

You have seen this interaction described by Powers' PCT model in Highlight 3.6, where the Input Function and related Perceptual Signal when compared with a Reference Signal may result in an Error Signal that influences the Output Function, including your muscle movements. In turn, this influences your environment with resulting Feedback that then influences your Input Function in a circular or closed loop manner. This circular signaling occurs so quickly that, in comparison to how slowly your muscles act, the causation that occurs in this circular loop can be thought of as occurring simultaneously.[4] Such rapid circular effects are why when you walk on an uneven surface, for example, you don't usually stumble and fall down; or when driving a car, you usually go at about the speed limit rather than going too fast or too slow; or when you are catching a ball, you often successfully catch that moving object. Similarly, when looking at your environment, you can easily see circular causation occurring between people, for example, during conversations, when two people dance, and when sports games are played.[5]

Internally, circular causation is a basis of life and behavior. As Powers has pointed out, "The root, the core, of the behavior of living systems is negative feedback control at every level of organization from RNA and DNA to a spinal reflex to a mental concept of physics. Negative feedback control is the basic principle of life."[6] And circular causation is an integral aspect of negative feedback control—an aspect that maintains stability in living systems and permits you and others to behave in ways that help you survive and experience what you want to experience.

When dealing with human behavior, it is important to realize that the illusion of linear causation occurs because the process often looks as if a stimulus actually causes a response. However, when negative feedback processes exist, as they do in all living creatures, the rapid circular loop control processes that occur in neurons are invisible to what one sees and experiences. This invisibility results in the stimulus-response illusion that occurs.[7] Therefore, more than simple observation is needed to understand what is happening when people do things.

Highlight 9.6 **Psychology and Circular, Closed-Loop, Causation**

"*Linear dynamic systems* behave nicely and can be easily analyzed, and that is why most scientists prefer to see the world in terms of linear systems.... Unfortunately, linear systems are idealizations rarely found in reality."

(Bossel, *Systems and Models*, 44)

"The classical cause-effect or input-output model of behavior breaks down when there is feedback from response to stimulus." "...when there is feedback from response to stimulus, such that there is a *closed loop* of cause and effect, conventional explanations which treat stimulus as cause and response as effect are no longer appropriate."

"The move to closed-loop psychology, when it happens, will be like starting psychology all over again, based on a new foundation...."

(Marken, *Mind Readings*, 61; and "You Say You Had a Revolution," 144)

Unfortunately, as the psychologist Richard Marken has pointed out, much of present-day psychology is based on linear thinking rather than closed-loop thinking and, as a result, is outdated.[8] This thinking is reflected in the types of experiments conducted and the statistical techniques that are routinely used—techniques such as multiple regression/correlation analysis, an approach that "is often (and properly) referred to as linear MRC" according to the statisticians Jacob Cohen and Patricia Cohen.[9] An example of the results of such thinking can be seen by looking at the diagram shown in Highlight 3.14. The flow of causation is shown to be one-way. Although helpful to thinking in many ways, such diagrams and the beliefs on which they are based are too simplistic and often misleading. Such thinking doesn't recognize how one's behavior affects that person's environment, which then affects the person's actions. For example, as you look at Highlight 3.14, you may realize that someone's actions can affect their opportunity for action, as well as their situational press, liability, emotional provocation from others, and even their assessment of costs and benefits of what they do.

Highlight 9.7 summarizes how these circular effects are a part of human nature.

Highlight 9.7 Human Nature According to Control Theory

"The control-theory version of human nature—or the nature of organisms, for that matter—can be put succinctly. Organisms control. Whatever we see them doing, at whatever level of analysis we prefer, we see them controlling, not reacting.

The old metatheory says that there is a one-way path through the organism, from cause to effect. The final effect is behavior. The new one says that there is a closed loop of action that has neither a beginning nor an end. The old concept says that behavior can be expressed as a function of independent variables in the environment. The new concept says that behavior is varied by the organism in order to control its own inputs. The old concept says that environments shape organisms. The new one says that organisms shape both themselves and their environments."

"According to control theory, it is the nature of human beings to control what happens to themselves as individuals and as a species. They are organized specifically as required to work in this way. That is a new conception of human nature."

(Powers, *Living Control Systems II*, 256-257)

To learn more about linear and circular loop thinking in psychology, see the articles by Richard Marken cited under "Further Reading" at the end of this chapter. And remember, things often are not as simple as they sometimes look—even to a sophisticated researcher![10]

Abstraction and Reality

As discussed in Chapter 7, the world out there and our perception of it are quite different. What we see, hear, feel, taste, smell, and otherwise perceive are constructions of our mind based on neural signals received from sensory neurons. These constructions are increasingly abstract, as the following sequence of events indicates:

a. **Something exists or happens in your environment.** For example, you walk down a dark street and a large man approaches you on the same sidewalk. Let's call this "the reality level."

b. **Your sensory organs are activated** and send impulses to your brain; you nonverbally see, hear, or otherwise sense something. Such silent sensations are abstractions since what is being perceived are actually nervous system impulses, not the real objects or events themselves.

What is sensed is also an abstraction since only some aspects of the object or event are sensed, not everything. These impulses and nonverbal perceptual signals are what we might call "low-level abstraction."[11]

c. **You may then recognize, interpret, classify, or describe what you are sensing** by using concepts and words that are as objective and close to the facts as possible. For example, you notice "A man walking towards me." In this way, words are used to describe things and events that are not words. That is, words are not the things or events they represent. They refer to things or events but are not the things or events themselves. Such words and descriptions, when they are as objective as possible, might be called a "somewhat higher-level abstraction."

d. **You may also give meaning to, think about, or otherwise make statements about your observations** of what is happening and use words or thoughts such as, "He may be a thief! He might try to rob and hurt me." Such inferred meanings and statements about words, such as when you see a man and think of him as a possible "thief" or "threat" is another level of abstraction, what might be called "a higher-level abstraction."

e. **In addition, you may also make use of or generate further inferences, interpretations, beliefs, theories, judgments, or conclusions**, about what you are thinking, such as "Dear God, please protect me from this thief!" or "Since God helps those who help themselves, maybe I should walk across the street to the other sidewalk." Such further inferences, beliefs, and theories, when verbalized, may be called "very high-level abstractions."

In other words, our sensations are about as close to reality as we can get. When we use words, we are operating at a more abstract level, although some words are closer to reality and less abstract than others. For example, the words "hard," "solid," "liquid," "red," "yellow," and "thirsty," "hungry," "sleepy," "itchy," and "eating" are less abstract than words such as "motivation," "attitudes," "willpower," "ego," and "social forces" that are often used to describe and explain what people do. Similarly, the word "apple" is less abstract than "fruit," which in turn is less abstract than "nutrition."[12]

In everyday speech and fields such as psychology, high-level abstract words are often used to describe and explain behavior—words that are far removed from reality as just described. These words are a form of shorthand that makes discussion possible. They are often useful when passing on information, coordinating action, and getting work done. In fact, all that we know are abstractions, verbal or otherwise.[13]

However, when we use theories and words that are highly abstract and far removed from reality, problems may occur. We are in danger of deluding ourselves with imaginary "just so stories." Such use of words, especially when unfalsifiable, may waste our time rather than help us to understand why behavior occurs and use that knowledge in practical ways. In addition, such words are often treated as real objects rather than the abstractions that they are, although, as mentioned, some are less abstract than others. For example, "body movements" can be observed without much inference. However, many words are simply inventions of someone's mind, conceived to help understand and explain what people do. Sometimes such invented words and ideas are helpful. Many times, they are not. For example, the idea of Descartes that the "the pineal gland is the seat of the soul" has not been very useful nor stood the test of time and scientific investigation. Similarly, the "ether" as a medium through which light travels has met the same fate. Whereas, other invented words such as "gravity," "atoms," and "molecules" have been helpful to scientists and engineers.

Likewise, when it comes to understanding what people do, some words and ideas have been more useful than others. For examples of words often used to describe and explain why behavior occurs, see Highlight 9.8. All the words shown are abstractions and not real things or real events in themselves, although some, such as "behavior," can be less abstract than others. At least "behavior" in the sense of body movements can be directly observed, whereas other words on the list, such as "attitudes," "Id," "repression," and "mind" are inferred or assumed based upon observations made, but are not directly observable.[14] Yet, even "behavior" has different meanings that can lead to endless confusion and fruitless discussion when the different meanings are mixed up during discussions and in professional writings. For example, Constantine Sandis points out that "behavior" can have at least three different meanings: (a) the physical movement of someone's body (such as she pushed a button, (b) the event or process of someone moving his or her body and doing something (she voted), or (c) what the person did, the result of the action, what was done (she broke with party policy).[15]

Highlight 9.8	Some Abstractions Used to Describe and Explain What People Do	
Action	Homeostasis	Regressed wish
Attention	Hopes	Reinforcement
Attitudes	Hysteria	Repression
Behavior	Id	Response
Behavior Setting	Incentives	Rewards
Beliefs	Intelligence	Roles
Body movements	Intentions	Satanic force
Cognitive maps	Interests	Schizophrenia
Conscientiousness	Intrinsic religiosity	Scripts
Culture	Knowledge	Self-control
Depression	Laziness	Self-esteem
Desires	Learning	Sexual regression
Drives	Love	Social forces
Ego	Memory	States of mind
Emotions	Mental states	Stimulus
Executive control	Mind	Stress
Expectations	Mood	Systems
Extraversion	Motivation	Tacit knowledge
Fears	Neurosis	Theory
Feedback	Norms	Time
Free will	Oedipus complex	Tradition
Glandular secretions	Perception	Traits
Goals	Perceptual signals	Unconscious
Group identity	Personality	Value
Habits	Purpose	Willpower
Happiness	Reflexes	

Another problem hinted at before is that many abstract words are treated as if they really exist. Such words are treated as real things that have certain causal powers. They have been "reified" or are a result of "objectification" as Alfred Korzybski calls it, "the confusion of words or verbal issues (memories, 'ideas'. ,) with objective, un-speak-able levels, such as objects, or experiences, or feelings." As he further explains, "If we objectify, we *forget*, or we *do not remember* that words *are not* the objects or feelings themselves, that the verbal levels are always different from the objective levels."[16]

As a result of reification and objectification, we read or hear statements such as "attitudes cause behavior" or "culture causes people to do certain things"—even though such statements are nonsense—treating the abstract word "attitudes" as a thing that causes behavior or the abstract word "culture" as another thing that causes people to behave in certain ways. Even professionals often seem unaware or forget that words do not directly cause behavior! Words are only abstractions that either describe things or events or are inferences about such things and events. As Alfred Korzybski said in an often-quoted statement, "A map is *not* the territory it represents."[17] Symbols of all kinds, including words, signs, and maps, are not the things to which they refer!

Korzybski captured these ideas by using the diagram shown in Highlight 9.9. The "Reality/Process Level" represents the atoms, molecules, and other characteristics of the real world—with each characteristic represented by a small black circular dot. We sense some of these characteristics as shown by the lines leading from the "Reality/Process Level" to the "Sensation/Object Level" of abstraction, but we do not sense other characteristics, represented by the small circular dots and lines not leading to the "Sensation/Object Level." Some characteristics of our sensations are then captured by descriptive, relatively-objective words at the "Label/Descriptive Word Level." However, other characteristics of our sensations are left out or otherwise not captured by words as shown by some dots at the "Sensation/Object Level" that are not connected to the "Label/Descriptive Word Level." For example, can you use words to fully describe sensations of "Love" that you may have? Methinks not!

Even more removed from what we experience at the "Label/Descriptive Word Level," we may make statements that are increasingly abstract, leave out some characteristics, and possibly introduce other characteristics concerning what is being sensed and expressed in words. These statements include the inferences we make about people and events, as well as theories developed that attempt to describe and explain behavior. Finally, as shown by Highlight 9.9, the arrow going back to the "Reality/Process Level" indicates how we project our assumptions, inferences, beliefs, and theories onto what we consider to be the real world. Doing so may then affect what we sense, perceive, and assume reality to be.[18]

Highlight 9.9 **Levels of Abstraction**

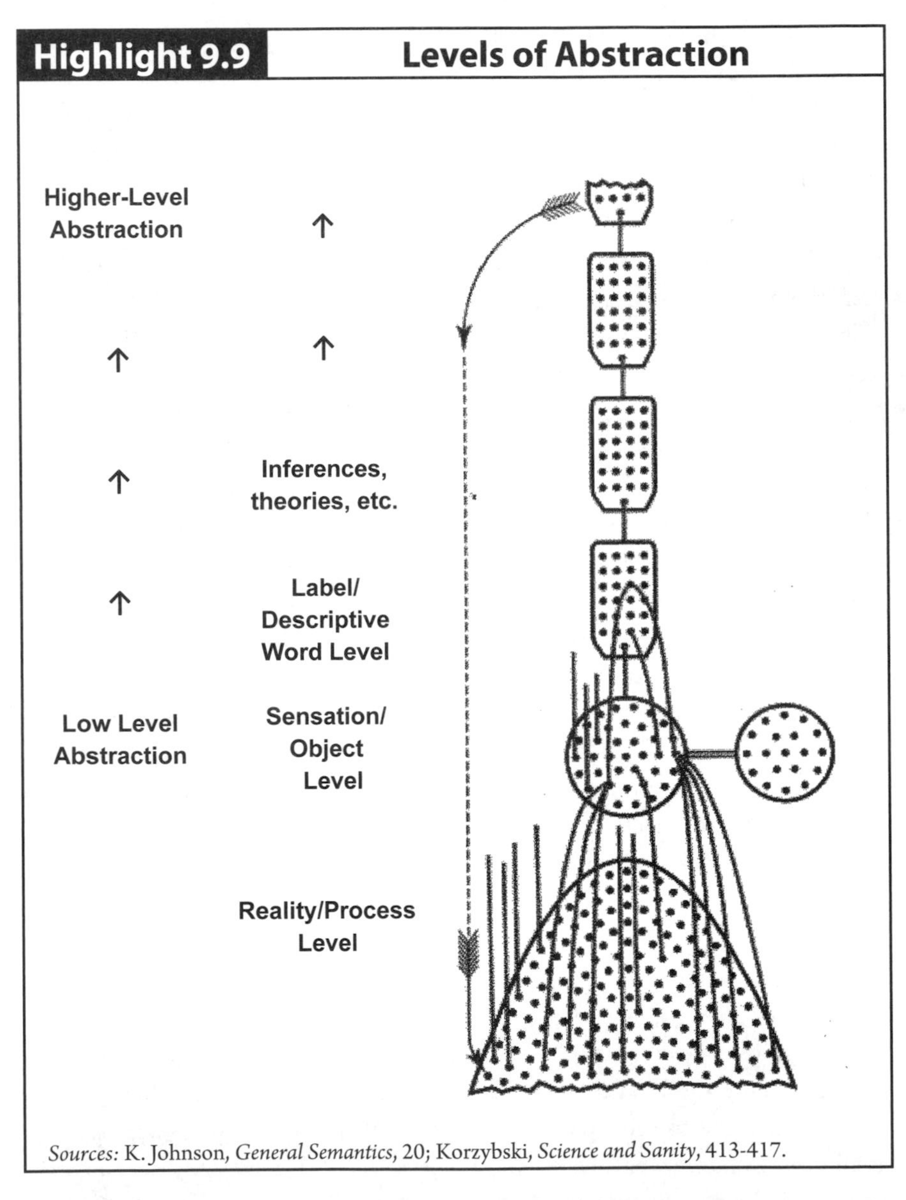

Sources: K. Johnson, *General Semantics*, 20; Korzybski, *Science and Sanity*, 413-417.

In fact, many psychologists have used highly abstract words so much that psychology dealing with the mind has been described as "words about words" with many such words being nothing more than "manufactured eloquence" according to Peter Munz.[19] That is, the words used are inferences or attempted descriptions about the body and its internal states and processes that are manufactured, made up abstractions—words such as happy, fear, rage, repression, and neurosis. The words interpret rather than describe neuronal events and sensations. As interpretations and abstractions, they do not really exist except in the thoughts of those who use the words. See Highlight 9.10 for more of Munz's thinking about psychology and its study of the mind.

Highlight 9.10 Psychology of the Mind and Manufactured Eloquence

"**psychology** *n.* **1.** the study of the mind and behavior."

(VandenBos, *APA Dictionary*, 2nd ed., 860)

"By psychology... we understand the study of states of mind—how they come into being, how they hang together, how they cease to exist, and how they influence one another."

"In psychology, where the events referred to by the eloquence have no existence apart from the eloquence—or better, where they exist but do not carry sufficient information to make any difference to the kind of eloquence that is manufactured about them—there can indeed be no distinction between literature and empirical psychology."

"... there is no subject matter. There are only constructions of eloquence and such constructions are unfalsifiable.... manufactured eloquence about silent neurons...."

(Munz, *Critique of Impure Reason*, 1, 109, 218)

Similarly, Stephen Turner, points out that theoretical concepts such as attitudes, culture, group consciousness, group mind, habits, ideology, implicit rules, norms, society, and tradition, are part of ordinary language. They are explanatory constructions, not collective objects that cause behavior. There is "nothing there" except summary descriptions.[20]

Other Shortcomings of Mainstream Psychology

Conflating Group and Individual Behavior

You are a unique person, different in many ways from anyone else. Although you are similar in some ways, you are also different.

A common shortcoming of psychologists and other social researchers is that many do not seem to recognize your uniqueness and the uniqueness of others. They overgeneralize the results of their research in ways that are often misleading. They often indicate that a pattern of behavior they find for a group of people applies to everyone. For example, they find that for some people, something (let's call it "X") is related to something else, such as what some people do ("Y"). They then make statements that seem to apply to everyone in the world such as, "People's automatic evaluations of their partners predict one of the most important outcomes of their lives—the trajectory of their marital satisfaction."[21]

When making such statements, many researchers do not indicate how many people exhibit this relationship. Does it hold for everyone or only some people? If some people, how many? 20 percent? 50 percent? 80 percent? Such scholars also do not usually indicate to which particular group in the world the relationship applies. Does it apply only to one group studied—such as American college students at one university during one particular year? Or to a sample of people from one European city? Or to nearly everyone alive today? Researchers often do not say so. They just make a blanket statement about "people" that seems to apply to everyone, or to most people, or to a broad group such as men, women, the poor, or disadvantaged.

Here are examples of other blanket statements:

- "We found that, when people observe that others violated a certain social norm or legitimate rule, they are more likely to violate other norms or rules, which causes disorder to spread."[22]
- "The female brain has tremendous unique aptitudes—outstanding verbal ability, the ability to connect deeply in friendship, a nearly psychic capacity to read faces and tone of voice for emotions and states of mind, and the ability to defuse conflict. All of this is hardwired into the brains of women. These are talents that women are born with that many men, frankly, are not."[23]

Many social scientists also use statistics about groups they have studied, to make statements about how an individual person behaves or will behave. As Highlight 9.11 indicates, such statements are "unsure" and may or may not be true. For example, even if 36 percent or 55 percent of people observed do something (as in the social norm violation study mentioned), that does not mean that the next person who comes along will do the same thing. If you think that he or she will, that's like thinking that since black men are more likely than white men to be put in prison within the United States, if you see a black man you should treat him or her as a criminal. Such stereotyping and overgeneralizing from a group to an individual or from some people studied to "everyone" in that group, as if everyone in a group is the same, is inappropriate. And yet many psychologists seem guilty of similar stereotyping and overgeneralization when they make blanket statements such as those just mentioned.[24]

Highlight 9.11 Individual and Group Behavior

"There is a deep gulf between statements about an identified individual and statements about a class [such as a group of people]. Such statements are of *different logical type*, and prediction from one to the other is always unsure."

(Bateson, *Mind and Nature*, 39)

The point is to be aware of such misleading statements. Not everything about human behavior that you may read or hear is correct. The reports of even well-conducted research may not apply to you—especially if such reports are overgeneralized, simplified too much, or stated in other misleading ways. Phenomena generated by statistical techniques applied to group data may not even exist in individual people! As Aaro Toomela, an Estonian psychologist, wrote, "It is possible to find regularities in behaviors that do not stem from a hidden mental cause or structure.... A phenomenon can instead be created, for example, when group data are analyzed.... A factor analysis...may create a phenomenon—such as a g-factor, neuroticism, extraversion, collectivism, etc.—that does not necessarily characterize mental organization of any person taken individually."[25]

Remember, *words are abstractions, not reality.*

Assuming that Relationships and Predictions Indicate Causality

Finally, many social scientists and laypeople observe a relationship between two things (X and Y) and then act as if X causes Y. Such causal thinking is faulty. The relationship found may be due to any number of different causes. For example, it may occur simply by chance, may be a result of something else that influences both X and Y, Y may in fact be causing X, or both may be influencing one another. In short, seeing a relationship tells nothing about what, if anything, causes the relationship.

Similarly, being able to predict something does not show that you know why the predicted event occurs. Predictions show relationships. They do not explain why those relationships occur. Do they occur just by chance, such as the strong relationship between the U.S. suicide rate and banana imports that I recall from my college days? Does a real causal relationship exist? Is there another explanation for the relationship? Predictions do not provide answers to such questions. Theory can.

Theory is often needed to understand the causes of behavior and presumed relationships between behavior and its causes. For example, a popular idea prevalent in psychology today is that a stimulus causes people to respond in certain ways. As you may recall from our introduction to autopoiesis in Chapter 1, this simple stimulus-response (S-R) causal relationship is incorrect in living creatures, yourself included! A stimulus does not cause a response. Instead, what you do depends on your internal organization and structure. The outside environment may trigger what you do. It may set into motion reactions on your part. But it does not cause you to do certain things or to behave in certain ways.

The apparent stimulus-response, cause and effect relationship is due to a "behavioral illusion." Powers and others have made this point quite clear.[26] To illustrate, as Marken explains in simple language, "The behavioral illusion is seen when an environmental variable appears to be acting on an organism to cause its responses. For example, we see this illusion when a person's head turns to see an attractive passer-by.... It looks like the passer-by causes the person's head to turn but the person is actually controlling the image of the passer-by, keeping it centered on the fovea."[27]

In short, people are not controlled by what they perceive. They are not controlled by stimulating their sensory neurons. What people do results from how their control systems operate, which in turn is based on the structure and organization of those systems. Powers explains that when feedback occurs, as shown by the PCT model, "One must characterize the organism's actions as behavioral control of stimulation and not stimulus control of behavior."[28]

These ideas should become clearer later when we discuss reinforcement and consequences under the topic of PCT as a "metatheory." Our previous discussion of circular versus linear causation in this chapter including Highlight 9.6 is also relevant to the S-R paradigm and related ideas of causality that attempt to explain the relationships between stimuli and what you and other people do.

Theories and Models—What are They?

Theories are invented descriptions of nature that are aimed at explaining, predicting, or otherwise helping us to understand the world around us.[29] Such statements should be judged by their usefulness, rather than considered to be true or false, since, as discussed before, abstractions such as theories are simplifications of the real world. Like any abstraction, theories and models do not capture all that there is to the event, process, object, or phenomenon being described. Many are so incomplete or abstract that they don't even try to predict behavior that will occur.[30] Partly as a result, many theories about behavior cannot be tested to see how closely they represent observation and experiment, rather than being "just-so stories" of someone's imagination.[31]

Highlight 9.12 A Few More Definitions

theory *n.* 1. a principle or body of interrelated principles that purports to explain or predict a number of interrelated phenomena.

principle *n.* ... in the empirical sciences, a statement of an established regularity....

model *n.* 1. a graphic, theoretical, or other type of representation of a concept or of basic behavioral or bodily processes (e.g., a disorder) that can be used for various investigative and demonstrative purposes, such as enhancing understanding of the concept or process, proposing hypotheses, showing relationships, or identifying epidemiological patterns.

empirical *adj.* 1. derived from or denoting experimentation or systematic observations as the basis for conclusion or determination, as opposed to speculative, theoretical, or exclusively reason-based approaches.... 2. based on experience.

generalization *n.* 1. the process of deriving a concept, judgment, principle, or theory from a limited number of specific cases and applying it more widely, often to an entire class of objects, events, or people.

(VandenBos, *APA Dictionary*, 2nd ed.)

The words "theories" and "models" mean the same thing to some people, but often "models" are more concrete representations of theories, sometimes represented as diagrams, flow charts, figures, equations, or physical structures. Highlight 3.6 is one example. Some models can be used to test the accuracy and usefulness of a theory by seeing how closely predictions based on the theory are linked to related observations and measurements. Modeling using computer-based simulation is sometimes used for this purpose and may be necessary when dealing with nonlinear, closed loop processes, whose outcomes are especially hard to predict. Such modeling, using computer programs, will behave the same way that the real thing behaves if the modeling and theory on which it is based is any good.[32]

Highlight 9.13 | **Helpful Models**

"The value of a well-designed simple model is that it helps one focus quickly on the most significant factors...."

(Houk, "Homeostasis and Control Principles," 256)

Unfortunately, many so-called "theories" in psychology are not theories at all. They are just descriptions and generalizations based on observations made. "Empirical generalizations" might be a better term to use with these mini-theories that number in the hundreds if not thousands. Most of these generalizations are limited by the fact that they are based on research conducted within only one culture or only one group, such as U.S. college students.[33]

Highlight 9.14 | **Theory and Models Must Be Testable**

"...theory must be testable; that is, a theory must ultimately be subject to empirical evaluation. The...constructs about the unconscious introduced by Freud, provide an example of an untestable theory."

(Jaccard and Jacoby, *Theory Construction*, 32)

"...let us remind ourselves...that every time we give a plausible explanation, chances are that someone else with some other theory can be just as plausible. What tests the claim is building a model."

(Runkel, *People as Living Things*, 180)

"The model, when expressed as a simulation, is the equivalent of a detailed and quantitative prediction of behavior that can be compared with real behavior."

(Powers, *Living Control Systems III*, 69)

"Any model should be investigated according to its merit with a view at the explanations and predictions it is able to provide. General criticism does not help, and the decision whether or not a model is suitable, exclusively rests with facts of observation and experiment. On the other hand, no model should be taken as conclusive; at best it is an approximation to be progressively worked out and corrected."

(von Bertalanffy, *General Systems Theory*, 183-184)

Good theory meets certain standards or criteria. For example, criteria for a unified theory of human behavior, according to Paul R. Lawrence and Nitin Nohria, are:

- "It should be multilevel, that is, able to work back and forth among all levels of analysis from individual to societal.
- "It should be consistent with most of the findings of the various social science disciplines, as well as with human biology.
- "It should be amenable to empirical testing.
- "It should be action-oriented—that is, practical (teachable and usable)—and, therefore as parsimonious as possible while still taking into account most of the principal features of human behavior.
- "It should be valid in different cultural settings."[34]

Most theories of human behavior do not meet these criteria.

Why Are Some Theories So Special?

Having looked closely at the question of why people do the things they do, some theories seem to stand out as being better for understanding what actually occurs. These theories are:

- Systems theory, including ecology, and
- Perceptual Control Theory (PCT) which is a subset of systems theory.

Why do these theories seem so special? Why are they emphasized in this book? The following sections explain why.

Systems Theory

As discussed before, systems theory helps us realize that life is not simple. We are composed of many systems both small and large. These systems consist of interrelated parts that affect one another. We are also part of larger systems that affect us and are affected by us and what we do. Ecological thinking, a form of systems thinking, helps us to recognize that we are part of broader systems and environments, that these environments influence us, and that we influence our environments in ways that also come back to influence us. "No man [or woman] is an island, entire of itself…" as John Donne wrote.[35]

Systems thinking also makes us aware that the characteristics of complex wholes such as a person like you cannot be reduced to the properties of their individual parts and that the relationships between parts are important when trying to understand behavior and other characteristics of systems. As a result, considering one part in isolation from others will not get you very far in understanding why you do the things you do. When studying a whole system such as you, the relationships between the parts of the system are important to consider, since the processes between them result in what the system does. Systems theory also leads us towards synthesizing thinking rather than the approach of many scholars who only focus upon a small specialized field of knowledge while not considering "the whole" of which their part is usually only one aspect. Concepts such as steady states, feedback loops, triggers, and multiple causality are all a part of systems theory thinking[36]—parts that are relevant to the concepts of autopoiesis, PCT, behavior, and survival discussed earlier in this book.

PCT is related to systems theory.[37] It focuses on one type of system—control systems. So let's move on to PCT and why it is emphasized in this book.

Perceptual Control Theory

A Major Paradigm Shift

Perceptual Control Theory seems to be a revolution in psychological thought—a revolution that those who understand it feel will have an impact upon psychology equivalent to the impact that Darwin's theory of evolution had on biology.[38] It represents a major shift in psychological thinking in several ways. It focuses on our control of what we sense and perceive, rather than on our actions. Its approach is not linear but deals with the circularity and recursiveness of causal processes that operate within and around us. It is comprehensive, seems closely linked to the biology of how our bodies actually operate, and recognizes the interactive nature of us and our environments as systems.

The idea that we control our perceptions, not our actions, runs counter to the popular views of most people who think about behavior. The traditional view of behavior is that it is the end of a process that starts in the environment or within the body. This leads to the illusion that "inputs" cause actions. The PCT view, as you should know by now, is that rather than being a result of input, behavior is your body's way of controlling its sensory inputs to keep them at their reference levels—levels required for you to live or that were otherwise established as preferences since your conception. What you control when you behave are called "controlled variables." These can be experimentally identified and verified by using what is called "The Test for the Controlled Variable."[39]

Reasons why PCT is so revolutionary and unsettling to traditional psychological thought are provided in Highlight 9.15 and the sections that follow.

Highlight 9.15 **PCT – A Paradigm Shift**

"The perspective provided by perceptual control theory....turns the traditional analysis of behavior on its head. Instead of the still-dominant view of seeing stimuli (both past and present) controlling responses (or perceptions controlling behavior, or environment instructing organism), the theory offers the unorthodox view of behavior as controlling perception through the organism's *control of its environment*."
(Cziko, *Without Miracles*, 111)

"It...appears that Powers and perhaps some others influenced by him are the only behavioral scientists who have been able to free themselves completely from the one-way cause-effect trap."
(Cziko, *The Things We Do*, 238)

"Science has a social as well as an intellectual aspect. Scientists are not stupid. They can look at an idea and quickly work out where it fits in with existing knowledge and where it doesn't. And scientists are all too human: when they see that the new idea means their life's work could end up mostly in the trash-can, their second reaction is simply to think 'That idea is obviously wrong.' That relieves the sinking feeling in the pit of the stomach that is the first reaction. Being wrong about something is unpleasant enough; being wrong about something one has worked hard to learn and has believed, taught, written about, and researched, is close to intolerable. All scientists of any talent have had that experience.... But those who recognize and embrace a revolution in science are the exception. Most scientists practice 'normal science' within a securely established—and well-defended—paradigm."
(Powers, "The World According to PCT," 4)

The Fact of Control

PCT is based upon the fact of control—a fact of living systems explained by Powers.[40] Control is essential to our lives as living, autopoietic beings. If we did not control our perceptions, we would not be alive. This may seem difficult to believe, so let's take a closer look at some concepts and questions about control.

- What is control?
- Why is control important?
- Why do we do things? Why do we act the way we do? Why do we control?

Highlight 9.16 helps to answer the first question, "What is control?" To control is to do something to bring your perception to a condition that matches a reference condition that you have. If you have an important goal, you do things to achieve that goal and then maintain that achievement until doing so is no longer relevant. If you have a physiological set point, you do things to achieve that set point, usually in order to live. For example, you may have a goal to see a certain movie, and so you go and see it unless another more important goal or something else interferes with you going. Similarly, you have a physiological set point for oxygen in your blood within a certain range, and so you breathe to obtain the oxygen you need to survive.

Highlight 9.16 — Definition of Control

> **Control:** "Achievement and maintenance of a preselected perceptual state in the controlling system, through actions on the environment that also cancel the effects of disturbances." "To control something is to act to bring it to a specified condition, and then maintain it close to that condition even if unpredictable external forces and changes in environmental properties tend to alter it."
>
> (Powers, *Behavior: The Control Of Perception*, 2nd ed., 296; *Living Control Systems III*, 1)

Why is control important? The answer is simple. It's important with respect to our physiological needs because it keeps us alive. It's also important for our other goals and references because reaching those goals helps keep us satisfied and happy—unless doing so conflicts with other goals that we have. When such conflict occurs, though, we can often do things to reduce or eliminate that conflict.[41]

Why do we do things? Why do we act the way we do? Why do we control? The bottom line is that we do things, behave, and control in order to achieve or maintain desired perceptual states. We do things to experience perceptions that match the references that we have for those perceptions. And, as explained before, we do such things to keep alive and to lead as satisfying and happy a life as we are able. Happiness is important to us. In fact, the Dalai Lama believes that "the very purpose of our life is to seek happiness."[42]

In short, the fact of control is undeniable. Perceptual Control Theory explains this fact of our existence very well.

A Metatheory

Perceptual Control Theory is also a "metatheory," applicable to the behavior of all organisms from bacteria to humans. It is also a framework that can incorporate other psychological perspectives and theories.[43]

Using Powers' PCT as a macro-basis for explanation, a number of other psychological theories and models are related to, explained by, or otherwise fall within the PCT framework. For example, some theories overlap with PCT since they are, in fact, based upon Powers' theory. Within the field of sociology, Affect Control Theory and Identity Control Theory are examples. Both focus upon special aspects of perception and behaviors that affect such perception. These aspects include one's sentiments, feelings and emotions about people, objects, and events in the case of Affect Control Theory. Identity Control Theory focuses on people's perceptions of themselves.[44]

Control theory has also been recognized as a metatheory by a number of scholars. Charles S. Carver and Michael F. Scheier, for example, state that, "The notion of reducing sensed discrepancies has a long history in social psychology, in topics such as behavioral conformity to norms (Ash, 1955) and cognitive consistency (Festinger, 1957; Heider, 1946; Lecky, 1945). The self-regulatory feedback loop, in effect, constitutes a meta-theory for such effects."[45]

According to several scholars, self-regulation theory also fits nicely into the cybernetic control theory framework of input, reference, comparator, output, and feedback.[46] Similarly, Carver points out how goals, which are of great interest in personality and social psychology, serve as reference values in control theory feedback loops.[47]

Building on Powers' reasoning, Carver and Scheier point out that control theory provides a useful framework and metatheory for personality, social, clinical and health psychology.[48] Warren Mansell and Timothy Carey show how Perceptual Control Theory provides a robust, integrative, theoretical framework for understanding the cause, maintenance, and treatment of psychological disorders using psychotherapy.[49] Robert Lord and Paul Levy similarly indicate "Control theory can be used as a meta-framework that helps to integrate and advance a substantial amount of motivational, cognitive, behavioural, and to some extent, affective theorizing."[50] Carey, Mansell, and Tai do likewise, indicating that PCT may be the first functional model to integrate biological, psychological, and social perspectives concerning life.[51] Howard Klein indicates the same usefulness of PCT for integrating explanations of work motivation literature and other perspectives.[52] Campion and Lord also state, "We believe that the capacity to assimilate varied literature is a major strength of control theory."[53] Huang and Bargh similarly look upon control theory as a useful framework for dealing with goal dynamics and for developing systems-based understandings of individual behavior.[54]

Webb, Sniehotta, and Michie have taken the further step of using a control theory model that draws upon Perceptual Control Theory to organize ten theories from social and health psychology used to predict and promote behavior change. They show how each of these theories relate to different aspects of the behavior change process as depicted by the control theory loop.[55] Richard Marken and Warren Mansell also indicate how perceptual control can be a unifying concept in psychology. They point out that a range of studies dealing with muscle control, animal behavior, personality and social psychology, organizational psychology, and developmental psychology, are all consistent with the idea that all behavior is a process of control—an idea central to PCT.[56]

In addition, the observations of other theories can often be better explained and understood by using PCT. For example, the statements, "Behavior is governed by its consequences," and "*Positive reinforcement* is the addition (or increasing) of a certain consequence that increases the likelihood of a certain behavior," are popular ideas and concepts of behavioral psychologists.[57] However, PCT explains why such phenomena seem to occur and why they do not occur. As Richard Marken points out, "Responses are not selected by consequences; rather, consequences are selected (and achieved against disturbances) by the organism."[58] This is evident when an error signal does not occur. For example, when an error signal does not occur, consequences such as food or money will have no effect, respectively, on the behavior of a rat who is not starved and hungry or on a millionaire who is offered $50 to do something that has no relevance to him. Similarly, consequences and popular "reinforcers," "incentives," and "rewards" will have no effect on your behavior if they are not relevant to you, your goals, needs, or error signals that you have or may have in the future. And so, thinking that behavior is governed by consequences is too simple.

Highlight 9.17 **The Reinforcement Illusion**

"A version of the behavioral illusion...can be described as the reinforcement illusion—the belief that an organism's behavior is controlled by environmental reinforcement."

(Cziko, *The Things We Do*, 225)

Although Powers recognized that "reinforcement" does occur, the interpretation offered by PCT is different from that of the behavioral tradition; it is more comprehensive and process-oriented rather than being merely descriptive. In PCT terms, "positive reinforcement" is anything that reduces error by its addition or increase, "negative reinforcement" is anything that reduces error by its removal or decrease, and "punishment" is anything that causes or increases error. For example, to a robber needing money to support a drug habit, stealing money is "positively reinforced" to a behaviorist, whereas a "PCTer" would interpret the stealing as reducing error caused by the robber's perception of not having enough money. Similarly, putting on a sweater removes the unpleasant sensation of feeling too cold, this being an example of "negative reinforcement" to a behaviorist. To the PCTer, though, putting on a sweater is a way of avoiding or reducing the error of feeling too cold. And although "punishment" to a behaviorist includes things like smacking a child or imposing a fine on someone, PCT explains that punishment is an example of error being caused by the smack, something a child doesn't like, and fines that most people with limited money try to avoid.

In short, PCT explains such ideas quite well. As Powers has stated, "For a thousand unconnected empirical generalizations based on superficial similarities among stimuli, I here substitute one general underlying principle: *control of input.*"[59]

For reasons such as these, Perceptual Control Theory is viewed as a metatheory and framework that can incorporate and explain many other psychological perspectives and theories. Such a role is important since, as Arthur W. Staats writes when discussing the disunity of psychology and its hundreds of different separate theories, "Broad overarching theories, or theories that make that claim, should be called on to address such problems."[60] PCT is such a broad overarching theory!

High Predictive Power

As indicated before, PCT is also impressive because it is testable and its predictions, in simple tests at least, have been nearly 100% accurate—setting perhaps what should be a new standard for psychological research.[61] See Highlight 9.18 for more details.

Highlight 9.18 The Accuracy of Prediction

"Once you know that the system is controlling a particular variable, you can predict its behavior with great accuracy."

(Marken, "Controlled Variables," 262)

"A crucial test of understanding is the ability or lack of ability to predict."

(Dawes, *House of Cards*, 77)

"If the behavior of a model does not match the behavior of the person within an error of a fraction of one percent, then the model should be improved or the theory should be given up.... PCT has at last enabled that kind of model to be built."

(Runkel, *People as Living Things*, 101)

"The correlations between the predictions of the [PCT] theoretical model and the individual's behavior... are consistently in excess of r = .955. That evidence persuades me of the merits of this model of purposive individual behavior as a basis upon which to proceed to develop a model of purposive social behavior."

(McPhail, *Myth of the Madding Crowd*, 205)

Concrete Concepts and Functions

In addition, the concepts and functions of PCT are close to reality and well-defined. For example, referring to neural impulses, Powers states that a *perceptual signal* is "The signal emitted by the input function of a system; an internal analog of some aspect of the environment" whereas the *input function* is "The portion of a system that receives signals or stimuli from outside the system, and generates a perceptual signal that is some function of the received signals or stimuli." These definitions and others are contained in his book, *Behavior: The Control of Perception* (1973; 2nd ed., 2005).

The point is that what Powers describes is much less abstract and more specific than the concepts of many other theories in psychology. Much of his model is also consistent with our own experience, such as when we do things to achieve our conscious goals.

Explanation of Different Types of Behavior

As mentioned in Chapter 3, PCT also includes a hierarchy of control systems, references, and perceptions. These range from the simplest level that controls perceptions of intensity signals up to perceptions of procedures, principles, and system concepts. As Carver and Scheier indicate in Highlight 9.19, this hierarchy of control systems allows one to explain phenomena as simple as moving a finger or blinking an eye, to matters as complex as going to church, honesty, and composing a song.

Highlight 9.19 **A Benefit of the PCT Hierarchical Approach**

"One of the important benefits conferred by this hierarchical sort of approach: It allows one to account successfully for the fact that exceedingly restricted and concrete behavioral acts (i.e., changes in levels of muscle tensions) are used to create behavioral events that are often so abstract as to seem completely unrelated to those concrete acts (e.g., writing an article, winning a tennis tournament, faithfully executing the office of president). Indeed, to the best of our knowledge, it seems to be the *only* approach that claims to provide such an account."

(Carver and Scheier, "Control Theory," 117)

Empirical Support from Practical Change Efforts

Successful change efforts help to show the validity and practical value of PCT. As Abraham and Michie have stated:

> "... if interventions including ... specific goal setting, self-monitoring of behavior, review of goals, and provision of performance feedback ... were found to be effective [in promoting a specified behavior, more than interventions that did not include these techniques], this would constitute an endorsement of control theory, whereas ineffectiveness among such interventions would imply that control theory was not a useful foundation for intervention design in that domain. Such analyses could identify important mediators of behavior change and highlight theories likely to be most useful to intervention designers ... ."[62]

In fact, several studies of interventions have shown that behavior change techniques consistent with control theory were significantly more effective than other techniques in changing weight loss, healthy eating, and physical activity behaviors. As reported, "Interventions that combined self-monitoring with at least one other technique derived from control theory were significantly more effective than the other interventions ... ."[63] and "Studies including more BCTs [behavioral change techniques] aimed at dietary change that are congruent with

Control Theory were associated with greater weight loss . . . ."[64]

A review of research concerning therapist-assisted interventions in treating adult problem behaviors such as depression, anxiety, and those concerning health also found that control theory components of self-regulation such as goal setting, self-monitoring, and related feedback had high "effect sizes."[65] In addition, an approach to psychotherapy based on PCT has proven to be effective in helping others who have distressing problems. The approach, called The Method of Levels (MOL), is a way of directing awareness to the appropriate place in the perceptual hierarchy where internal conflict can be reorganized.[66] Psychological problems, distress, and internal conflict occurs, according to PCT, when we are not able to control some of our perceptions. Timothy A. Carey, a clinical psychologist and developer of MOL, writes, "Psychological distress is the experiencing of enduring unwanted perceptual activity," that "Internal perceptual conflict occurs when two incompatible perceptions are to be controlled at the same time," and that "For conflict to be resolved, goals need to be changed, not behaviors," such goals being those that are setting incompatible lower level goals.[67] An example is when you want to do one thing, but you also want to do something else, such as be a nice polite person, but tell someone off, or not offend your conservative parents but you really want to live with a special friend of the opposite sex. The MOL approach to resolving such conflicts seems to work well in helping people resolve their conflicts and resulting distress.[68] Highlight 9.20 indicates how it works.

Highlight 9.20 | **The Method of Levels**

"The focus of MOL . . . is to *restore* control . . . by resolving conflicts. This is accomplished by helping psychologically distressed people shift their attention away from the symptomatic distress to higher level goals and purposes . . . . When attention reaches the higher levels at which the problem is maintained, . . . reorganization allows a shift in the way that lower order goals are balanced and regulated by the lower order systems."

(Mansell and Carey, "A Century of Psychology," 349)

Therefore, the use of Perceptual Control Theory as a framework for the design of behavior change interventions and related research seems appropriate. That is a reason why Chapter 11, which focuses on how to change your behavior, includes activities central to PCT. These elements include clarifying your goals, focusing on doing things to achieve those goals, self-monitoring, and consciously obtaining feedback to perceive how well you are doing.

The basic point of this section, though, is that components of control theory used in behavior change interventions have been shown to be effective in helping to achieve behavior change—further supporting PCT and its usefulness.

An Individual Focus

PCT also focuses on the behavior and perceptions of individuals, such as you, rather than on the behavior of groups of people. And so, in comparison to many theories that are based on data about groups, PCT can be used to help understand your individual behavior, rather than the average behavior of groups whose general tendencies may or may not apply to you.

PCT Meets the Criteria of Good Theory

Finally, you may recall that the criteria for a unified theory of behavior were mentioned when discussing models and theories. PCT meets those criteria. That is, it is multilevel, ranging in explanatory power from processes at the level of bacteria, to the nervous subsystem level, to the individual level where phenomena such as goal-related behavior and reinforcement are explained, to the societal level, as research and publications by sociologists such as Clark McPhail (1991) and Kent McClelland and Thomas Fararo (2006) indicate.[69] PCT seems consistent with findings of the social sciences and biology and helps to explain many of those findings, which is a reason why it is called a metatheory. It is amenable to empirical testing, and predictions based upon its modeling are extremely accurate. It is usable and consistent with action-oriented behavior change intervention efforts including psychotherapy. It is also parsimonious—as Highlight 9.21 indicates. Admittedly, though, it takes several years before most people fully understand its processes and implications, which is perhaps a reason why it is not more popular now and why, sometimes, scholars who write about it have only a superficial understanding of its essence and meaning. And lastly, in terms of Lawrence and Nohria's criteria for a unified theory of human behavior, PCT does seem quite applicable to different cultural settings, having a strong base of research and scholarship in places as diverse as the U.S.A., Europe, and China.[70]

Highlight 9.21	**PCT is a Parsimonious Metatheory**

"For a thousand unconnected empirical generalizations based on superficial similarities among stimuli, I here substitute one general underlying principle: *control of input*."

(Powers, "A Cybernetic Model for Research," 164)

Because PCT meets the criteria of good theory so well, it has been used in this book to help you understand why you do the things you do. As one scholar, Peter Lipton, has written, "Better explanations explain more types of phenomena, explain them with greater precision, provide more information about underlying causal mechanisms, unify apparently disparate phenomena, or simplify our overall picture of the world."[71] As the previous sections of this chapter indicate, PCT does these things quite well.

Preview of the Next Chapter

The next chapter will help you to apply what you have learned by showing how to systematically analyze your behavior. By doing so, you should be able to understand even better than now why you do the things you do.

Further Reading

For an easy readable introduction to systems theory, look at:

- Donella H. Meadows, *Thinking in Systems: A Primer* (White River Junction, VT: Chelsea Green Publishing, 2008).

For more comprehensive coverage of systems theory as applied to living systems, see:

- Fritjof Capra and Pier Luigi Luisi, *The Systems View of Life: A Unifying Vision* (Cambridge: Cambridge University Press, 2014).

For a review of how people in the social sciences became aware of the feedback loop concept, what they have accomplished with it, and what they perceive its significance and limitations to be, see:

- George P. Richardson, *Feedback Thought in Social Science and Systems*

Theory (Waltham, MA: Pegasus Communications, 1999).

A book whose goal is to establish the literal reality of negative feedback control as the basic organizing principle of human behavior is:

- William T. Powers, *Living Control Systems III: The Fact of Control* (Bloomfield, NJ: Benchmark Publications, 2008). Includes computer simulations.

A good book of readings about Perceptual Control Theory containing 34 papers and samples from 26 books, downloadable without charge as a pdf file from www.living control system.com:

- Dag Forssell, ed., *Perceptual Control Theory: An Overview of the Third Grand Theory in Psychology* (Hayward, CA: Living Control Systems Publishing, 2016).

For discussions of psychology's apparent fixation on open-loop, linear causation and why PCT is a revolutionary paradigm shift, see:

- Richard S. Marken, "You Say You Had a Revolution: Methodological Foundations of Closed-Loop Psychology," *Review of General Psychology* 13, no. 2 (2009): 137-145.
- Marken, Richard S. and Warren Mansell, "Perceptual Control as a Unifying Concept in Psychology," *Review of General Psychology* 17, no. 2 (2013): 190-195.

An easily read introduction to Alfred Korzybski's ideas concerning levels of abstraction is:

- Alfred Korzybski, *Selections from Science and Sanity: An Introduction to Non-Aristotelian Systems and General Semantics*, 2nd ed. (Fort Worth, TX: Institute of General Semantics, 2010).

For a discussion of psychology's manufactured eloquence concerning the mind, see:

- Peter Munz, *Critique of Impure Reason: An Essay on Neurons, Somatic Markers, and Consciousness* (Westport, CT: Praeger, 1999).

Other critiques of mainstream psychology include:

- Aaro Toomela, "Variables in Psychology: A Critique of Quantitative Psychology," *Integrative Psychological and Behavioral Science* 42, no. 3

(September 2008): 245-265. A scholarly view of the variables and statistical analysis techniques used by psychologists.

- Robyn M. Dawes, *House of Cards: Psychology and Psychotherapy Built on Myth* (New York: Free Press, 1994). Focuses mainly on the therapy profession and its assumptions.
- Thomas Teo, *The Critique of Psychology: From Kant to Postcolonial Theory* (New York: Springer, 2005). A scholarly, historic view.
- Jane Osden, "Some Problems with Social Cognition Models: A Pragmatic and Conceptual Analysis," *Health Psychology* 22, no. 4 (2003): 424-428. A scholarly, critical article.

Endnotes

1. "All purposeful behavior may be considered to require negative feed-back" according to Rosenblueth, Weiner, and Bigelow, "Behavior, Purpose, and Teleology," 19.

2. As Philip Zimbardo has pointed out, "Traditional analyses [of behavior] by most people, including those in legal, religious, and medical institutions, focus on the actor as the sole causal agent. Consequently, they minimize or disregard the impact of situational variables and systemic determinants that shape the behavioral outcomes and transform actors"—from Zimbardo, *The Lucifer Effect*, 445.

3. Laszlo, *Systems View of the World*, 6.

4. Runkel, *Casting Nets and Testing Specimens*, 93.

5. Clark, *Being There*, 165.

6. Powers, *Living Control Systems III*, xi-xii.

7. A better understanding of how stimuli can affect behavior and the basis of behavioral illusions can be found in Powers, "Quantitative Analysis of Purposive Systems"; Marken, "The Nature of Behavior," reproduced in Marken, *Mind Readings*; and Cziko, *The Things We Do*, 87-89, 127, 174.

8. Marken, "You Say You Had a Revolution." Henry H. Yin also points this out in his chapter "Restoring Purpose in Behavior."

9. Cohen and Cohen, *Applied Multiple Regression/Correlation Analysis*, 11. The quote shown is repeated in the 3rd edition of the same book (p. 9) where it is also stated that "MRC, ANOVA, and ANCOVA are each special cases of the *general linear model* in mathematical statistics" (p. 4). Heise also writes in *Causal Analysis*, "*Linear relations are presumed in many statistical analyses and in all the analyses in this book*" (p. 92).

10. It should also be noted that the if-then of logic and logical sequences that guides much scientific thinking is different from the if-then of cause and effect, because the if-then of logic is (a) timeless, (b) unable to deal with circular recursive circuits without generating paradoxes, and (c) an incomplete model of causality. For details see Bateson, *Mind and Nature*, 18-19, 54-55.

11. Damasio calls such sensations "somatic markers" and Munz calls them "the buzzes we sense." For details see Damasio, *Descartes' Error*, 173, and Munz, *Critique of Impure Reason*, 14.

12. Burgoon, Henderson, and Markman, "There Are Many Ways," 501-520.

13. Hayakawa and Hayakawa, *Language in Thought and Action*, 86-87, 93.

14. However, even a term such as behavior can have multiple meanings. To some people, the term refers to body movements, but even they concede that behavior can be motionless such as pressing one's hand against a door to keep it closed, standing still, keeping quiet, or staring at something intensely as Sandis points out in *The Things We Do*, 6. But the term can also refer to events or processes like eating, swimming, and fighting and what someone did, such as celebrated, tried to break a record, or killed a person. Even when we restrict ourselves to behavior involving outward bodily movement, different people may have different things in mind, such as the mere movement of the body, the event or process of an agent moving his or her body, or what the agent did as Sandis explains in a fairly deep philosophical discussion (see Sandis, *Things We Do*, 6-11).

15. Ibid. A less philosophical article about the many meanings of behavior is that of Levitis, Lidicker, and Freund, "Behavioural Biologists Do Not Agree."

16. Korzybski, *Science and Sanity*, 417.

17. Ibid., 58.

18. Korzybski, *Selections from Science and Sanity*, 136-153; K. Johnson, *General Semantics*, 20; and Korzybski, *Science and Sanity*, 25-27.

19. Munz, *Critique of Impure Reason*, 5, 218-219.

20. Turner, *Social Theory of Practices*, and *Brains/Practices/Relativism*. Chinese tradition and philosophy also recognized the difference between the way things are and the names, words, and distinctions used to describe and represent those things. For example, see de Bary and Brown, *Sources of Chinese Tradition*, vol. 1, 380-386.

21. McNulty and others, "Though They May Be Unaware," 1120.

22. Keizer, Lindenberg, and Steg, "The Spreading of Disorder"—an article which, to its credit, does indicate percentages of people affected, but neglects to indicate that only a minority of subjects actually fit the conclusions in five out of six field studies conducted, with those apparently influenced being 36 percent, 55 percent, 28 percent, 28 percent, 12 percent, and 14 percent respectively in those six studies—as Powers pointed out on the

Control Systems Group Network in a posting, "Re: Statistics in Psychology," December 23, 2008. As Powers indicated, at best we could say that "some people" and at worst that "a few people" are more likely to be affected, at least in Groningen, Netherlands, where the studies were conducted.

23. Brizendine, *The Female Brain*, 8.

24. Runkel also discusses such stereotyping in his book, *People as Living Things*, 186-187.

25. Toomela, "Variables in Psychology," 248.

26. Powers, *Selected Papers*, a reprint of "Quantitative Analysis"; also Marken, *Mind Readings*, 24-27.

27. Marken, *More Mind Readings*, 56.

28. Powers, "A Feedback Model for Behavior," 47, 58.

29. Reynolds, *A Primer in Theory Construction*, 934.

30. Weinstein, "Testing Four Competing Theorics," 326.

31. For example, a criticism of the Health Belief Model, Protection Motivation Theory, the Theory of Planned Behavior, and the Theory of Reasoned Action is that they are essentially flawed because they cannot be tested—according to Ogden, "Problems with Social Cognition Models." A "just-so story" describes an unverifiable and unfalsifiable narrative explanation for a trait or behavior of humans or other animals according to http;//en.wikipedia.org/wiki/Just-so story, December 29, 2011.

32. Runkel, *Casting Nets*, 112, 132.

33. Jahoda, "Critical Comments," 357.

34. Lawrence and Nohria, *Driven*, 14.

35. Seldes, *The Great Quotations*, 209, with spelling changed to modern usage.

36. At least these concepts are a part of the thinking of Bertalanffy in his classic book, *General Systems Theory*.

37. Bertalanffy, *General Systems Theory*, 17, considers cybernetics to be a part of a general theory of systems. Also, since PCT is based on classical negative feedback control theory that is the basis for cybernetics according to Powers, *Living Control Systems III*, 2, and since Powers' work is a combination of what George P. Richardson calls the cybernetic and servomechanism threads in his book, *Feedback Thought*, 240, PCT can be considered to be either a part of, influenced by, or otherwise related to systems theory.

38. Powers, "Quantitative Analysis," 434; Marken, "You Say You Had a Revolution," 137-145.

39. This test is described by Powers, "The Nature of Robots," and more recently by Runkel, *People as Living Things*, 76-78, and Marken, *Doing Research on Purpose*. "The

Test," as it is sometimes called, involves doing research on an individual person rather than by obtaining statistics about groups of people as is usually done in psychology. Such a group approach is not very suitable for determining how humans function as Runkel points out in *Casting Nets.*

40. For example, Powers, *Living Control Systems III,* focuses on this fact and includes computer programs on compact disks to demonstrate how negative control systems operate.

41. The Method of Levels, described a bit later in this chapter, is one way to reduce such conflict. Powers discusses the basis of conflict in both editions of his book, *Behavior: The Control of Perception.*

42. Lama and Cutler, *The Art of Happiness,* 1.

43. Carver and Scheier, "Control Theory"; Hyland, "Control Theory and Psychology"; Johnson, Chang, and Lord, "Moving from Cognition to Behavior"; Klein, "Control Theory"; Lord and Levy, "Moving from Cognition to Action"; Sheeran and Webb, "From Goals to Action"; Webb, Sniehotta, and Michie, "Using Theories of Behaviour Change"; and Carey, Mansell, and Tai, "A Biopsychosocial Model."

44. For introductions to these theories see Heise, *Expressive Order,* and Burke and Stets, *Identity Theory.* Robinson, "Control Theories in Sociology," mentions other authors and work and credits Powers "landmark work" of 1973 as initiating this surge of theoretical development.

45. Carver and Scheier, "Self-Regulation of Action," 6. Also see Carver and Scheier, "Cybernetic Control Processes," 38, where they point out that a metatheory is a general way of conceptualizing how complex systems work and that control theory does so by explaining how behavior occurs rather than what behavior occurs.

46. MacKenzie, Mezo, and Francis, "A Conceptual Framework"; Carver and Scheier, *Self-Regulation of Behavior.*

47. Carver, "Self-Regulation of Action," 14.

48. Carver and Scheier, "Control Theory."

49. Mansell, "Control Theory and Psychopathology"; Carey, "Exposure and Reorganization."

50. Lord and Levy, "Moving from Cognition," 98.

51. Carey, Mansell, and Tai, "A Biopsychosocial Model."

52. Klein, "Control Theory," 38-39, 40.

53. Campion and Lord, "A Control Systems Conceptualization," 270.

54. Huang and Bargh, "The Selfish Goal," 154, 162-164.

55. Webb, Sniehotta, and Michie, "Using Theories of Behaviour Change." Theories

reviewed include Goal-Setting Theory, the Model of Action Phases, the Strength Model of Self-Control, Protection Motivation Theory, the Theory of Planned Behavior, the Health Belief Model, the Elaboration Likelihood Model, the Prototype Willingness Model, and Social Cognitive Theory.

56. Marken and Mansell, "Perceptual Control as a Unifying Concept."

57. Ramnerö and Törneke, *The ABCs of Human Behavior*, 82, 83.

58. Marken, *Mind Readings*, 27.

59. Powers, "Quantitative Analysis of Purposive Systems," 164. For more information about how PCT acts as a framework that can incorporate other theories and views, see Marken, *More Mind Readings*, 23-35, who explains why stimulus-response views, reinforcement views, and cognitive views, although apparently different ways of describing behavior, "can now be seen as legitimate ways of describing different aspects of one phenomenon—control. Each is just a different way of describing what an organism must do to keep its sensory inputs at their reference values," 34.

60. Staats, "Unifying Psychology," 11.

61. In this respect, Perceptual Control Theory does especially well when compared to other theories of behavior, since modeling based on PCT has resulted in predictions of 94 percent to nearly 100 percent accuracy (i.e., correlations of .971 to .998). For details see Marken, "Perceptual Organization of Behavior"; Bourbon, "Invitation to the Dance"; Bourbon, "On the Accuracy"; and also "Empirical Evidence for Perceptual Control Theory," accessed from http://www.pctweb.org/EmpiricalEvidencePCT.pdf, on February 29, 2012.

62. Abraham and Michie, "Taxonomy of Behavior Change Techniques," 381-382.

63. Michie and others, "Effective Techniques in Healthy Eating," 690. When discussing control theory they refer to Carver and Scheier whose thinking was obviously influenced by William T. Powers. In fact, Carver and Scheier acknowledge Powers' generous review of a draft of the material concerning control theory and the hierarchy of control structures contained in their book, *Attention and Self-Regulation*, and refer to Powers frequently in that publication(see ix, 395). In the study reported by Michie and others, intervention techniques studied that the authors associated with control theory were "prompt intention formation," "prompt specific goal setting," "prompt self-monitoring of behavior," "provide feedback on performance," and "prompt review of behavioral goals." In a later publication, however, Michie et al., in "A Refined Taxonomy of Behaviour Change," 1485, state that "intention formation" should actually be labeled "Goal Setting," and they did so in their revised taxonomy of behavior change techniques (BCTs).

64. Dombrowski and others, "Identifying Active Ingredients," 7.

65. Febbraro and Clum, "Meta-Analytic Investigation, 153-158."

66. Carey, *The Method of Levels,* 143.

67. Ibid., 29, 44, 57.

68. For more information about MOL and its grounding in PCT theory, besides Carey's *Method of Levels,* see Carey and Mullan, "Evaluating the Method of Levels"; Mansell and Carey, "A Century of Psychology"; Mansell, "Control Theory and Psychopathology"; Carey, "Exposure and Reorganization"; and Carey, Mansell, and Tai, *Principles-Based Counselling.*

69. The website http://www.livingcontrolsystems.com/files/applic_pct.html indicates that PCT is applicable to fields as diverse as management, physiology, neurology, medicine, sociology, psychology, resolving psychological distress, education, learning, personal relationships, infant development, robotics, cybernetics, artificial intelligence and philosophy, and evolution.

70. A good place to link up with this network of scholars and research is at http://www.iapct.org, the website of the Control Systems Group: International Association for Perceptual Control Theory. To join the web discussion group, send an e-mail to listserv@listserv.illinois.edu with the words "subscribe csgnet" or go to https://lists.illinois.edu/lists/arc/csgnet and sign up.

71. P. Lipton, "Causation and Explanation," 629.

CHAPTER 10

HOW TO ANALYZE YOUR BEHAVIOR

The Big Picture

Having read this far, you now have a good idea of the many internal and external influences on your behavior. You also understand why a systems perspective and Perceptual Control Theory (PCT) are highlighted in this book. This chapter will help you use what you have learned to analyze and better understand your behavior. It does this by showing how to focus on a behavior of interest to you—something that you are doing now or have recently done—and answering some questions based on the PCT control loop concerning that behavior.

A Slightly Expanded PCT Model

As Chapter 9 indicates, a systems theory perspective and Perceptual Control Theory have influenced this book. They provide a useful framework for understanding and analyzing human behavior.[1] With this framework in mind, I will now help you integrate what you have learned so far. I'll do this by using three diagrams that reflect you as the large control system which you are, containing many smaller hierarchically organized control systems as discussed in Chapter 3. Based on the diagrams, we will then go over a series of questions that you can use to better understand why you do the things you do.

The first diagram, shown in the previous chapter as Highlight 9.1, shows the hierarchy of systems that affect you during your lifespan. These range from the molecular and cellular level up to your broader environment that includes the economy and distant weather. As suggested by Highlight 9.1, both you and these systems change during your life. For example, the friends you had as a child are probably different from those you have now. Even if some are the same people, they have changed over time. Similarly, you, your cells, and your internal systems have also changed.

The second diagram, contained in Highlight 10.1, shows you as a control system, but in more detail than you have seen before. As you can see, the familiar functions and concepts of your nervous system as modeled by PCT are shown—starting with your senses, the input function, and perceptual signals that are compared to reference signals. Differences that occur or are expected to occur then produce error signals that stimulate your output function and resulting behavior that affects the environment around you. That environment, in turn, continues to continuously stimulate your senses and input function, and so on around the nervous system loop until you either reduce your error signals, avoid anticipated error signals, or do something else.

Highlight 10.1 **An Expanded PCT Model**

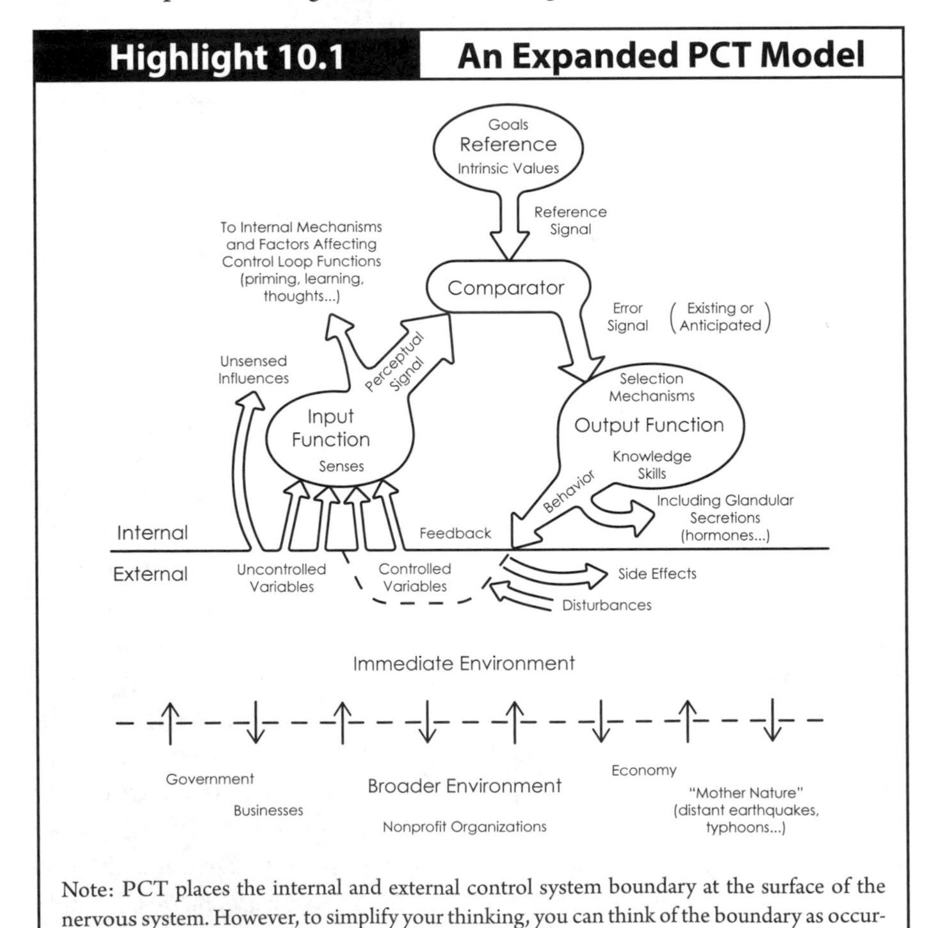

Note: PCT places the internal and external control system boundary at the surface of the nervous system. However, to simplify your thinking, you can think of the boundary as occurring at the surface of your skin.

A few more details are shown by Highlight 10.1 than in the basic PCT model presented in Chapter 3. For example, Highlight 10.1 indicates that you may sense and perceive things that you are not trying to control, such as the moon at night or, often, what someone else is doing. These other things sensed are called "uncontrolled variables" in contrast to the "controlled variables" that you perceive and try to control. Uncontrolled variables may influence you in important ways such as by priming your neural networks, by attracting your attention, and by affecting what you learn such as during "observational learning" as discussed in Chapter 4.

Highlight 10.1 also shows that some variables that you sense may affect various "Internal Mechanisms and Factors"—again priming your neural networks and influencing processes such as learning, memory, and thought. In addition, as indicated by the Output Function, your "Behavior" may be of several types. That is, your behavior includes not only muscle movements but also glandular secretions such as hormones released into your bloodstream as well as tears, breast milk if you are a lactating mother, and possibly pheromones.[2]

Highlight 10.2 indicates some internal mechanisms and factors that may affect your perceptions, references, and what you do.[3] While you may or may not be conscious of these mechanisms and factors, they do affect your control systems and how they operate. For example:

- Your Input Functions and resulting perceptual signals may be affected by prior learning, expectations, hormones, priming, drugs, and other factors such as low glucose levels due to recent activity.
- Your Output Functions and resulting behavior may similarly be affected by what you have learned before, such as your skills, beliefs, and knowledge of what works; by your habitual way of doing things based on similar perceptions experienced in the past; by recall of that learning; by priming some responses; by how confident you are that you can successfully achieve your goal; and by fatigue, drugs, and other factors.
- Your Reference Values and Signals may also be affected by earlier perceptions that prime certain references, previous learning, realizing that what you are doing is not working and that you may have unrealistic reference values, drugs such as alcohol, and by other factors.
- Similarly, your Environment may affect you by means of influences on your body that are sometimes not directly sensed by your nervous

system, such as carbon monoxide gas; by means of sensed but uncontrolled variables, like the death of a close friend or relative; and by means of sensed variables that influence internal mechanisms and factors that are either a part your control systems or affect how they function, such as when you learn how to do something that you are not yet controlling.

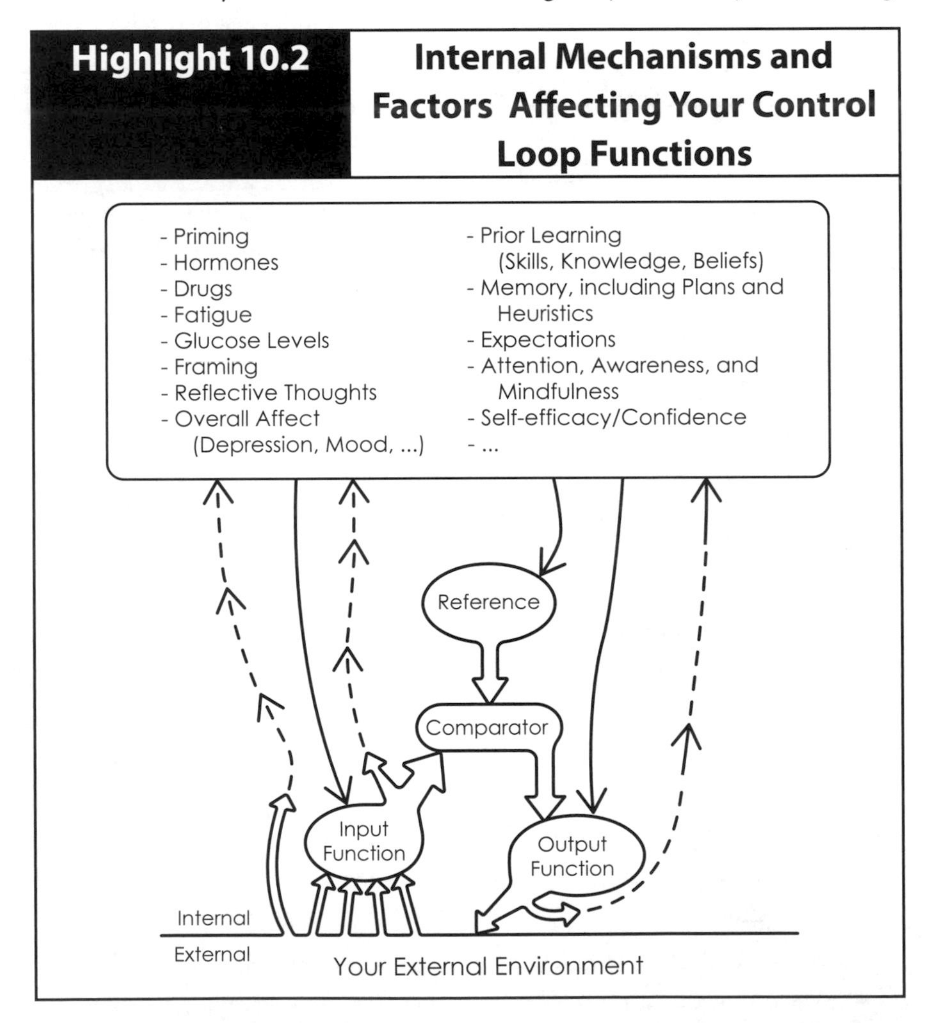

The phenomenon of "feedforward" shows such factors in operation. Feedforward is a term used by some social scientists and engineers to refer to an action based on anticipation or prediction that occurs before an error signal is actually experienced. For example, when you go outside in cold weather, you often put on a coat or jacket before you feel cold. You perceive that it is cold by having heard or read a local weather forecast, by looking at an outdoor thermometer, or perhaps by seeing snow outside. This stimulates or primes your memory concerning the reference that *if* it's cold outside and you decide to go out, *then* you should put on warm clothing such as a coat or jacket.[4] That reference when activated results in an error signal related to the fact that it is cold outside and you have not yet put on enough warm clothing. As a result, you put on a coat, jacket, or other warm clothing before going out.

Another example is when you look for car keys, glasses, or something else that you have recently put down. If you remember where you put them, you simply go there, see them, and pick them up. If you have misplaced them because you were distracted by something else when you put them down, you look around until you eventually find them—perhaps first looking in places where you usually put them and then, if not found, searching around more randomly. What you do is another indication of the effect of memory on your output function. If you remember where the object is, you go straight to it. If you don't remember where it is, you end up looking around until you find it.

Similarly, perhaps a cue or reminder you have written primes your reference to buy milk so that when you pass the milk counter at the supermarket, you put a container in your shopping cart. Otherwise, without that priming, you may forget to buy milk.

The other mechanisms and factors shown in Highlight 10.2—like hormones, drugs, and fatigue—also affect the operation of your control loops. They are either a part of those control loops or are general descriptions of things that affect the functioning of those loops.[5] Some are elements of the "Sub-Personal Level" shown in Highlight 9.1, such as neural linkages associated with memory and learning. Some are elements at the "Cell and Molecular Level" such as hormones, neurotransmitters, and drugs. Others are descriptive concepts at the "Personal/Individual Level" such as expectations, thoughts, and memory. Although at different functional and logical levels, these mechanisms and factors are shown in Highlight 10.2 since they can help us think about the many internal influences on the operations of our control loops and resulting behavior.[6]

In a similar way, we can think about external environmental factors and mechanisms that influence our bodies and perceptions. However, since these were discussed extensively in Chapters 7, 8, and 9, you can look at those chapters again if you wish to remind yourself of influences that are external to your body.

Using the Expanded PCT Model to Analyze Your Behavior

What to Do

We'll now apply what you have learned to help you systematically analyze and better understand your behavior. As an aid, we'll use the slightly expanded PCT model shown in Highlight 10.1 and Highlight 10.2 to:

- Go around the PCT control loop one or more times, and then
- Conduct additional analyses of internal and external factors to further your understanding, if you want to do so.

To do this, answer the questions shown in the following sections. The first section shows how to analyze your current behavior. The section after it shows how to analyze what you have done in the past.

How to Analyze Your Current Behavior

The First Major Questions

If you want to analyze something that you are doing now, try to answer the questions shown below in ***bold type***. We'll call these the "major questions." As you will see, each major question is followed by other questions which will help you think about possible answers to the major question. The numbers shown in Highlight 10.3 correspond to the number at the beginning of each major question to remind you about what part of a control loop you are considering when you answer that question.

1. What are you doing?

Start with a behavior that is of interest to you. In other words, what are you doing now that you want to analyze and better understand? Make sure that you focus on an observable behavior—something that you are doing with your body. If you are not sure how to describe that behavior, think of this about yourself: "If another person was observing me, what would that person say I am doing?"

For example, a person might observe that you clean and straighten up your house or apartment every day. Or that you smoke a pack of cigarettes a day. Or contribute a lot of money to charity. Or go to the gym and exercise there regularly. Or go to bed late and don't get enough sleep.

Highlight 10.3 Parts of the Expanded PCT Loop to Consider When Analyzing Your Behavior

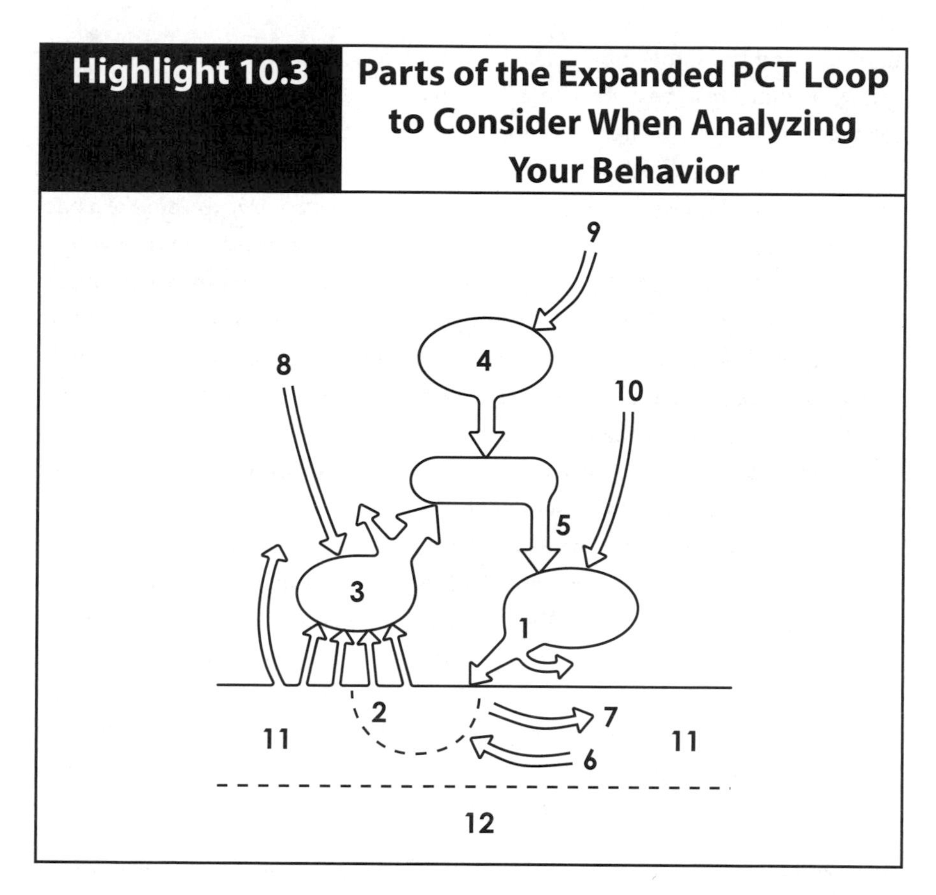

Actually, what you might choose to analyze is almost endless. The behavior you choose to focus on is up to you. But to answer the question, remember that the behavior you choose should be observable, either by someone else or by you.

If you are still not sure how to answer this question, don't worry! I'll give more examples later as illustrations of answers to this and the other major questions that follow. I'll also provide a guide that you can copy and write your answers on. For now, just keep on reading to get a sense of what to do.

2. *What are you trying to control?*

As you may recall, you behave in order to control things so that your perceptions match references that you have for them. These things or "controlled variables" as they are sometimes called may be the appearance of an actual object, a process, event, situation, condition, person, or even a feeling—anything that may vary and has some importance to you.[7] And so, when answering this major question, indicate as best you can what you are trying to control with the behavior that you described when answering major question 1 above.

For example, if you clean and straighten up where you live every day, are you trying to control how your place looks if people stop by? Or if you smoke a pack of cigarettes a day, do you feel uneasy, upset, or bad in some way, and so you are trying to control and reduce that unpleasant feeling? Or do you smoke to control how you look, and smoking looks cool and helps you fit in with your friends who smoke? If you give money to charity, are you trying to control your taxes by obtaining a tax reduction? Or reduce suffering in the world? Or publicize yourself and look better to others? Or control how you feel about yourself, because you feel better after giving to charity. Or control several of these things at once?

In other words, what present or future feeling, thing, event, process, or condition are you trying to change, avoid, or continue? For instance, is there a feeling that you are having or might have, such as hunger, thirst, a desire for something, or something that is wrong that you want to change? Is something occurring in an undesirable way that you want to change? Are you trying to control something that is happening now? Or are you trying to control something that will happen in the future? Is it what someone is doing or might do? Are you trying to change or maintain the condition of your body in some way such as your knowledge, skills, or appearance?

3(a). What was your perception of what you are dealing with **before** *you started doing what you are doing?*

Before you started doing what you are doing, you had a real or imagined perception of what the thing was like that your behavior is directed towards. Maybe something was wrong that you wanted to change. Or maybe you wanted to avoid something from happening. Or perhaps you wanted some desirable condition to continue.

Whatever the situation, what was your perception of what you are trying to control *before* you started doing the behavior you are analyzing? Or to put it more technically, what was your perception of the controlled variable or variables you are trying to control before you started doing what you are doing?

For example, "The place was a mess!" Or, "I felt like an outsider because my friends were smoking and I wasn't." Or, "People were suffering and needed help." Or, "I was fat."

3(b). What is your perception of what you are dealing with **now?**

Now that you are doing something, your behavior may have had some impact. Or perhaps what you are doing has had no effect that you can notice. Or your behavior may be keeping some situation from changing. In any case, when answering this question, think about what you are perceiving now that you have taken action.

In other words, what is your perception of the controlled variable or variables, now, since doing what you are doing? Has anything changed? How is what you are doing useful to you? What desirable result is your behavior producing or maintaining?

For example, "My place looks really nice." Or, "I smoke like my friends do and feel like a part of the group." Or, "I'm making donations, helped people in need, and feel good about it too." Or, "I look slim and trim now."

4. What is your reference for what you are doing?

According to PCT, you do things so that your perceptions match your important goals and other references as much as possible. And so, an important part of analysis is to identify what reference or references are linked to the behavior you are analyzing. In other words, to what desirable end is your behavior directed? What are you trying to achieve? What value, goal, desire, or other reference is guiding your action?

To help you identify the reference involved, you might look at Highlight 3.4 ("Words Indicating References") to remind yourself of some of the many different references you have. You may even have more than one important reference influencing your actions, so if you identify more than one, that's OK! When answering this major question, though, try to identify the highest reference, the ultimate "Why?" that is driving your behavior and related perceptions of what you are trying to control.

For example, a person keeping a clean and tidy place might have references like, "I don't want to look like a slob to others, and besides, 'Cleanliness is next to Godliness.'" A smoker might have references like, "I want to be like my friends" and "Smoking calms me down" and "I always smoke after I eat." A person giving donations might think, "The Government is too big and I don't want to support it by paying taxes and giving it my money" or "The Lord Jesus said, 'It is more blessed to give than to receive.'"

5(a). What error was occurring or anticipated **before** *you began doing what you are doing?*

What was the difference between your perception and reference before you took action? Was the difference important? What seemed wrong before you started taking action? How did you feel about what you perceived and what you desired? Such differences and feelings can be noted as your answer to this question. If more than one set of differences and errors occurred, that's OK too. Each can be noted, with emphasis given to the most important one.

If no error was occurring, did you start doing things to prevent an error from happening? For example, maybe you started taking action to maintain your perception of conditions that you desire or prefer. And so, did you start taking action to avoid the occurrence of a future error? If so, and an error was anticipated, you can note that as your answer to the major question.

For example, "My place is a mess and my friends are coming over!" Or, "My friends smoke and I don't; I'm different." Or, in the case of going to the gym, "I was overweight, had too much fat, and often strained my back."

5(b). What is the error like **now?**

What difference is your action making? Is the error less now? To what extent was the error reduced? How do you feel now after doing what you are doing ? Is the error being avoided, reduced, or prevented from increasing? Whatever the case, answer the question accordingly.

For example, "My place looks neat and clean." Or, "My friends and I are alike and do things together." Or, "My weight is under control, I look good, and haven't had back pain for years."

6. What else in the environment is affecting what you are trying to control?

Think about other things that are affecting the variable you are trying to control. What else in the environment supports or reduces the effectiveness of what you are doing? What are those things or events? Try to identify them.

For example, "Those who live with me keep making a mess around the place." Or, "My friends often offer me a cigarette." Or, "I regularly received requests for donations in the mail and on the phone."

7. What side effects is your behavior having?

In addition to what you are trying to control, your behavior also affects other things. What are these side effects? Being aware of them, especially if they are important, may help you decide whether to continue or change what you are doing. For example, if a side effect ends up harming other people or may harm you later, knowing that may help you to change your behavior. If the side effects are positive or not important, you may wish to continue behaving as you are, if doing so helps reduce error that you are experiencing.

For example, "Those who live with me know that they don't have to clean up, because I'll do it." Or "I can't go up a flight of steps without coughing; my breath smells; and, the money I spend on cigarettes can't be used to buy other things that I want."

Other Possible Major Questions

After considering the seven major questions just discussed, if you wish to probe further, you might consider other influences on you, your control loop functions, and on what you are trying to control. Answers to the following questions will help you do this.

8. What influences are there on your input function and resulting perceptions?

Many things affect your perception. These range from internal factors such as your goals, needs, interests, feelings, emotions, political or religious biases, stereotypes that you hold, expectations, the familiarity of what you are sensing, knowledge, skills, and your body's condition—are you tired, drunk, or sick. As discussed in Chapter 7, aspects of what you observe can also affect what you notice and perceive—including movement, novelty, and context, such as hearing a noise in bed on a dark night, or at work, or during a party.

For example, if your goal is to make the floor as clean as possible, you may notice more dirt and spots than you would otherwise perceive. If you know that you are going to sunbath and swim at the beach during an upcoming vacation, you may notice your body shape, fat, and weight more than during other times of the year.

9. What influences are there on the references you are using?

Is anything priming the reference you are using? Does a cue remind you of your reference? Have you set a goal to do something? If so, why? Is your reference influenced by other people? Have you been persuaded by them, or otherwise adopted a goal of others that seems like the right thing to do? Are you guided by expected behavior, norms, or standards—such as religious commandments, a job description, standard operating manuals, policies or laws? Are you reminding yourself of a reference, such as by reflecting on "What would Jesus (or another respected person) do?" Is your reference a sub-goal required to achieve a higher goal, principle, or broader plan that you have?

For example, if you see someone else smoking or a pack of cigarettes, does this prime your goal to smoke a cigarette? If you are in a church or hospital, do you know that you are expected not to smoke there? If a friend asks you to go to the gym, does this activate your reference to go?

10. What internal influences are there on what you are doing?

Are you tired, fatigued, hungry, thirsty, excited, sexually aroused, feeling confident, and/or drugged? Do you have the skills required to do something else? Do you think that there is something positive about doing what you are doing rather than doing something else to reach your goal?

For example, if you are sick, are you as likely to clean and straighten up your house or apartment? If you are nervous or anxious, do you reach for a cigarette? If you are drunk, do you go to the gym? If another goal you have is more important than exercising at the gym, will you go?

11. What other immediate environmental conditions are affecting you and what you are doing?

Are other people around and does this affect your behavior; for example, are you copying or imitating what they do? Is what you are doing prompted by the environment you are in such as by cues, signs, or other people? Do others encourage or support you in doing what you are doing? Are you affected by external conditions such as advertising or the Premack principle whereby you are required to do one thing before you can do something else such as having to enter your PIN number or password before withdrawing money from an ATM? Is your behavior guided by a checklist, implementation plan, action plan, decision tree, policies, or architecture of the context you are in?

For example, if your friends keep their places neat and clean, you might do this also. If you see an advertisement for cigarettes, you may be tempted to smoke. If you go to church, you will probably give a donation when the collection plate or basket is passed around.

12. What broader environmental influences are there on what you are doing and its effects?

Is what you are doing a result of the broader world? As discussed in Chapter 8, are things affecting what you are doing such as cultural norms; your status in society; the role you are in; kinship ties; your religion; national events such as the state of the economy, political disturbances, or a war being waged; the country you are in; government laws and policies; the school curriculum that determined what you were taught and expected to learn; what broader organizations are doing, such as your employer, advertising, politics, mass media; diseases that are broadly occurring like AIDS, Ebola, measles; and the state of technology and science including the medical and communications systems.

For example, business people may place advertisements on television, because they remind people to buy and use their products or do certain things. Religious policies may encourage you to behave in certain ways, such as to donate or tithe 10 percent of your income to charity or the church. Similarly, broad environmental influences that may affect how late you go to bed and how much sleep you get include technologies such as electric lights that provide light at night; computers and smart phones that may capture your attention; TV programs and movies that go on late into the night; evening and night shifts at work; and traditions such as New Year's Eve celebrations, midnight mass on Christmas Eve, bachelor and bachelorette parties before weddings, and Late Night Breakfasts or other events before examinations at many colleges and universities.

Where We Are and What's to Come

By asking and answering the major questions just presented, you should be able to better understand why you are doing something that you are doing now. The next section indicates how to analyze something that you have done in the past—a recent or earlier behavior. After that, to make the major questions and possible answers more concrete, we'll look at two examples of behavior analyses.

Analysis of Recent or Earlier Behavior

Let's consider now how you can analyze something that you have done in the past. The farther back in time you go, the hazier your memory will be of events and conditions that affected what you did. As a result, if you are analyzing a past behavior, it is best to focus on something that you have done recently rather than something that you did a long time ago.

To analyze things that you have done recently or did earlier use the same questions in the previous section, but ask them in the past tense. For example, instead of asking, "***1. What are you doing?***" for a past behavior, ask "***1. What did you do?***" Instead of asking, "***2. What are you trying to control?***" to analyze a past behavior, ask "***2. What were you trying to control?***" Do the same for the other major questions. Just ask them in a way that refers to the past.

If all of this seems a little abstract, the next sections contain concrete examples from my own life that indicate how it is done. The first example is an analysis of something that I recently did. The second example is an analysis of something that I do almost every day.

Example 1: An Analysis of Something I Did Today

Here's something that I did just before writing this section: I took my car in today for a 30,000 mile service although it had only 29,888 miles on it. Obviously, I took it to the garage for a routine service, but early—that is, 112 miles (179 kilometers) before the service was due.

To analyze and understand why I did this, here are my answers to the major questions indicated before:

1. What did you do?

Answer: I made a 30,000 mile service appointment for my car two days ago and took the car today to have it serviced at the garage of the dealer where I bought it.

2. What were you trying to control?

Answer: I was mainly controlling for proper functioning and safety of the car—which is to be used by my son to take his wife and child to a distant airport in a few days to go on an international flight.

3(a). What was your perception of what you were dealing with* before *you did what you did?

Answer: My son needed to use the car in a few days. The proper functioning and safety of the car was slightly in doubt since the odometer reading was approaching the recommended service mileage, the service had not been carried out, and a snowstorm was predicted for tomorrow and the next day, just before my son needed the car.

3(b). What was your perception of what you were dealing with* after *you did what you did?

The 30,000 mile service was successfully carried out. The car should be safe to drive a long distance.

4. What was your reference for what you did?

Answer: My references were keeping my family safe—by making sure the car was in good shape for my son's trip; the 30,000-mile service recommendation by the manufacturer which helps ensure the proper functioning and safety of the car; and doing the service today to avoid problems and delays in having the car serviced should it snow 5 to 10 inches (13 to 25 cm.) as predicted during tomorrow and the next day.

5(a). What error was occurring or anticipated* before *you began doing what you are doing?

Answer: The car had not been serviced and the error, the difference between my perception and reference level, was important, because I wanted my son's family to arrive safely at the airport for the trip, and the car was approaching the recommended mileage for service. I was uncomfortable with the situation before the car was serviced.

***5(b). What is the error like* now?**

Answer: The error was eliminated. I am satisfied with the roadworthiness of the car.

6. What else in the environment affected what you were trying to control?

Answer: The dealer was able to schedule the appointment for today when I called two days ago. He was also open today and had the manpower and parts to do the required service. The weather and roads were fine. I had money in the bank to pay. My personal obligations, schedule, and finances enabled me the freedom to do what was required—and so I could drive to the dealer and pay for the service.

7. What side effects did your behavior have?

Answer: After paying the dealer for the service, I now have less money and the dealer has more. I was not able to spend my time doing other things. The dealer recognizes me as a regular customer. I contributed to air pollution by driving to and from the dealership.[8]

Additional probing questions and answers are as follows:

8. What influences were there on your input function and resulting perceptions?

Answer: I can't think of any other internal influences besides the references noted before. My son's request to use the car seems to be the major external influence.

9. What influences were there on the references you used?

Answer: Influences included my belief in doing routine service on a car according to the manufacturer's recommendations, a general value of caring for my family and helping avoid harm to them, and perhaps my general adherence to the belief that "it is better to be safe than sorry."

10. What internal influences were there on what you did?

Answer: Besides the beliefs and value just mentioned, my memory helped me drive along the familiar route that I drove to the dealer where the service was to be performed. Also, my tendency to schedule appointments with the dealer from whom I bought the car and my belief that the dealer knows the car well and can provide good service on it, influenced selection of my behavior to call that dealer.

11. What immediate environmental conditions affected you and what you did?

Answer: My daughter-in-law wanted to take the flight together with her 6-month old son and had made reservations to do so. My son asked if he could use my car since his car was old and somewhat unreliable. The weather forecast that I heard and read about affected the timing of the appointment that I requested before the snowstorm. No one else was available to drive the car to the dealer for service. I also had time to take the car for service, having no other conflicting obligations.

12. What broader environmental influences were there on your behavior and its effects?

Answer: For one thing, a wedding of my daughter-in-law's brother was to take place in a foreign country and that is why she scheduled a flight for the coming weekend. The manufacturer's suggested service schedule was an important factor as was the heavy snow that was predicted to come our way.

Summary of this analysis:

- **My Behavior:** I made a 30,000 mile service appointment for my car two days ago, took the car to the dealer today, and had it serviced.
- **The Controlled Variable:** I was mainly controlling for the proper functioning and safety of my car during my son's trip.
- **Initial Perceptions:** My son needed the car in a few days to drive his family a long way to the airport. I perceived that the odometer reading of the car was approaching the recommended service mileage and the service had not been carried out. A snowstorm would occur two days before the trip.
- **References:** My major references were a value for keeping my family safe, the service recommendation of the manufacturer, and a desire to ensure that the car was serviced and safe to drive when needed by my son.
- **Error Signals:** The car had not been serviced and I was uncomfortable with that fact.
- **Side Effects**: I was not able to do other things while the car was being serviced. The dealership obtained money for the service provided, and I have less money.
- **Environmental and Other Internal Influences:** A number of other factors influenced what I did including the weather prediction, my belief in the dealer's competence, my ability to pay for the service, the dealer's ability to schedule an appointment on the day requested, and my memory of how to go to the dealer.

- **My Perception Now:** The car has been serviced and I am satisfied that it is safe for my son to drive to the airport on Saturday with his wife and child.

Example 2: An Analysis of Something That I Do Every Day

Here's an analysis of something that I do routinely, nearly every day: "Eat a Low-Carb Breakfast." Since I am dealing with a current, routine behavior, here are my answers to the major questions listed before.

1. What are you doing?

Answer: Eating a low-carbohydrate breakfast every day, whenever possible.

2. What are you trying to control?

Answer: My intake of carbohydrates and, ultimately, my weight, excess fat, appearance, and health.

3(a). What was your perception of what you are dealing with before you started doing what you are doing?

Answer: I was more than 20 pounds overweight. I looked fat, especially around my abdomen.

3(b). What is your perception of what you are dealing with now?

Answer: I am at my desired weight and no longer see much fat around my abdomen.

4. What is your reference for what you are doing?

Answer: One reference is the Atkins Diet Plan that advocates controlling the carbohydrates that one eats and eating within two hours after arising from bed in the morning in order to keep one's weight under control.[9] Another reference is to have a Body Mass Index (BMI) of less than 25—which is the cutoff for being overweight and having an increased risk of various health problems according to the World Health Organization and national health agencies.[10] My culture also generally values a trim body without much fat, and I have adopted that value.

5(a). What error was occurring or anticipated <u>before</u> you began doing what you are doing?

Answer: I could see a lot of fat around my abdomen and felt that I was overweight. My Body Mass Index was over 25. I also felt hungry if I didn't eat within an hour or two after rising in the morning.

5(b). What is the error like <u>now</u>?

Answer: The error was reduced and is now nonexistent. My Body Mass Index is below 25 and I don't appear to be excessively fat.

6. What else in the environment is affecting what you are trying to control?

Answer: My wife prepares low-carb meals for me. When I prepare breakfast myself, low-carb foods such as eggs and sausages are available in the house to eat. When eating outside my home, low-carb foods are sometimes not easily available.

7. What side effects is your behavior having?

Answer: My wife avoids giving me high-carb foods. My friends and relatives are sometimes surprised that I don't eat much of the high-carb foods that they prepare such as potatoes, rice, and delicious deserts. The Atkins Nutritionals company benefits because I buy some of their products. I also seem to serve as a positive example to my physician about effects of the Atkins diet.

Additional probing questions and answers are these:

8. What influences are there on your input function and resulting perceptions?

Answer: My routine of not working until I have had breakfast sometimes draws my attention to the fact that I have not yet eaten. Sometimes, so do my feelings of hunger. Looking at our household clocks also helps make me aware that it is time to eat.

9. What influences are there on the references you are using?

Answer: The broader culture of which I am a part has standards that have affected my references for being too fat. These standards are both aesthetic and based on medical evidence concerning obesity and good health. I have accepted and adopted those standards. In addition, the perceptions and awareness mentioned in my answer to Question 8, remind me of my references about eating a low-carb breakfast. In other words, after getting up in the morning, thoughts about my routine of eating before working, occasional feelings of hunger, and seeing the time, all help to prime my references.

10. What internal influences are there on what you are doing?

Answer: My knowledge, beliefs, memory, occasional hunger pangs, and habitual behavior linked to cues that I notice in the morning, all seem to influence what I do. For example, I know what foods are low-carb and high-carb and choose what I eat accordingly. Similarly, I believe that eating a low-carb diet can help me maintain my desired weight level because my experience with such a diet has helped me to lose weight and maintain my desired weight level.

11. What immediate environmental conditions are affecting you and what you are doing?

Answer: Mainly the cues indicated before such as time shown by household clocks, the availability of low-carb food to eat, my wife's routine preparation of low-carb meals, and a lack of distractions and demands on my time that would interfere with eating breakfast, seem to be the main immediate environmental conditions that affect what I do.

12. What broader environmental influences are there on your behavior and its effects?

Answer: The following seem to affect what I am doing: my culture's value on a slim trim body, a monthly retirement income that permits me to be choosey about what I eat, research-based health information indicating that being overweight can lead to health problems, and national food distribution systems that make low-carb foods conveniently available for purchase.

Summary of this analysis:

- **My Behavior:** I eat a low-carb breakfast every morning when possible.
- **The Controlled Variable:** My intake of carbohydrates—and, ultimately, my weight.
- **Initial Perceptions:** I was overweight and looked fat.
- **References:** The Atkins Diet Plan and internationally recognized Body Mass Index (BMI) recommendations that indicate normal weight.

- **Error Signals:** Occur when I eat too many carbohydrates, when my BMI is 25 or higher, and when I have not eaten breakfast and feel hungry. However, at present there are no error signals.
- **Side Effects**: Mainly, my wife now prepares low-carb breakfasts.
- **Environmental and Other Internal Influences:** Broader cultural standards concerning looking fat, the BMI and related medical information about weight and good health, Atkins Diet Plan books in my home, time shown by household clocks, availability of low-carb foods at home, my financial ability to buy low-carb foods, my knowledge about low-carb foods and the Atkins Diet Plan, my beliefs in Atkins Diet effectiveness and BMI-related medical recommendations, and occasional hunger pangs.[11]
- **My Perception Now:** I am not overweight. I ate a low-carb breakfast this morning. There are no related error signals occurring.

Guides for Your Use

By now, after reading this and previous chapters, you should be able to analyze your behavior as we have discussed. That is, if you want to understand why you do the things you do, you can use the questions and approach introduced to guide your efforts. However, with highly important but difficult to analyze behavior, you may need the help of others. The next chapter indicates how to obtain such help.

Before moving to that chapter, though, note that Appendices 2 and 3 contain guides that you can use to help you analyze and better understand the things you do. The guides contain the major questions shown before as well as spaces for your answers. Use Appendix 2 if you wish to analyze something that you are doing nowadays. Similarly, Appendix 3 can guide you if you want to analyze something that you did in the past.

As mentioned before, these approaches to analysis are grounded in Perceptual Control Theory. Highlight 10.4 speaks to that point well by the person who first developed that theory.

Highlight 10.4	Understanding Behavior
"Understanding behavior means knowing what perceptions are being controlled, how they are being controlled, and why." (Powers and others, *Perceptual Control Theory: A Model*, 7)	

When doing an analysis, however, here's a point to keep in mind: *You may not be able to identify all of the important factors that affect your behavior.* Some factors may be hidden mental processes of which you are not aware. For example, some knowledge, called "implicit knowledge" or "tacit knowledge" is stored without awareness and not easily described by using words. Even experts often cannot fully explain the knowledge and rules that they use when doing things.[12] As one definition of tacit knowledge states, "tacit knowledge . . . is informally acquired rather than explicitly taught.... It is stored without awareness and therefore is not easily articulated. Many everyday skills are of this kind, such as the ability to recognize faces or to speak one's native language."[13] Similarly, the neural signals of your control loops are silent and not directly accessible to your conscious awareness.

So, don't feel bad if you can't fully identify the reasons behind what you do. Just do the best you can using this book and the guidelines provided. You should do very well!

Analyzing Someone Else's Behavior

Other people operate the same way you do—at least as far as their control systems are concerned. Like you, others also do things to control their perceptions so they match their own references. Being aware of that will help you generally understand the behavior of other people. However, understanding the specific things that others do is often more difficult. Doing so requires finding out what a person's internal references are as well as their perceptions and other hidden influences, since you cannot observe such things directly.

One key to understanding why a person does something is to determine what variable that the person is controlling or trying to control. This can be determined in several ways. One way is to talk with the person and ask questions. For example, ask, "Why did you do that?" or "Why are you doing that?" If the person is honest and aware enough, you may be able to find out what he or she is controlling fairly quickly.

For something that they are doing now, you might also ask things such as "What do you want to change?" or "What doesn't seem right, that you want to change?" or "What is it that you want to keep the same?" If what the person is doing is aimed at achieving something in the future, you might ask, "What do you want to achieve by doing what you are doing?" Answers to such questions may also indicate important references the person has that influence his or her behavior.

Another way to identify a controlled variable is to change things to determine what the person is trying to keep the same, achieve in the future, or change in a certain way. In other words, guess what the person is trying to control, then interfere with what you think the person wants to achieve, and see if the person tries to counteract or neutralize what you are doing. If the person reacts and tries to reduce or eliminate the impact of what you are doing, you are affecting something that the person is trying to control. If the person does not act, you may not be dealing with one of his or her controlled variables. This way, of course, is more complicated and difficult than simply talking with the person. But sometimes this approach is needed until a pattern becomes clear enough for you to determine what variable is being controlled.

An alternate and easier way of identifying a controlled variable is to ask the person to imagine a situation in which something is changed, and then ask how would she or he then feel—the "something changed" being a description of what you think is the controlled variable.[14] Strong positive or negative feelings can help you identify what the person is controlling or wants to control.

Once you know what variable is being controlled, you will have taken an important step towards understanding why the person is doing what he or she is doing. And you will have answered major question #2, "What are you trying to control?" Hopefully you will use language that the person understands, since the word "control" may not be a word that the person is familiar with or knows how to consider.

In turn, you might try to answer the other major questions indicated in the previous sections. If you do, talk with the person in a way that he or she can understand. Avoid words such as references, control, and error signals if the person is not acquainted with Perceptual Control Theory. Instead, use such words as goals, objectives, values, or others shown in Highlight 3.4 rather than "references." Or ask what the person hopes for and wants. Similarly, instead of the word control you might use the word influence, such as by asking, "What are you trying to influence?" Instead of "error signal" you might use the idea of "difference," such as asking, "How do you feel when you see a difference between your goal and what exists now?"

You might also try to talk with others who know the person well to obtain or verify answers to some of the questions.

Answering the major questions in this way, if you can, will help you further understand why the person behaves as he or she does. Such an approach requires effort on your part—effort that is worth spending if an understanding of the person's behavior is especially important. In such cases, doing these things is better than just assuming or guessing why someone did something and then not doing anything to see if your assumption or guess is correct!

The major difference between analyzing your own behavior and analyzing someone else's is that you are privy to your own internal self—your thoughts, feelings, memories, intentions, many of your references, and perceptions. This helps you answer the questions more completely and (hopefully) more accurately.[15] When it comes to someone else, you know none of these things unless you talk with the person or are intimately acquainted with him or her. Even then, some of the information you obtain or infer may be spotty or misrepresented.

Even though other people function as you do—having perceptions, references, error signals, and hierarchies of control systems—obtaining details about such private knowledge for an accurate analysis requires that you make a special effort to obtain that unobservable information. Analyzing and understanding someone else's behavior may be more difficult and less accurate than doing so for yourself. But still, if the other person is honest and aware enough, you should be able to get a fairly good idea of why he or she behaves in that way or behaved like that. However, recognize that others may only tell plausible stories, rather than give knowledgeable answers—especially if they haven't read this book, are not familiar with some of its concepts, and are not able to recognize important internal and external influences on their behavior.

Preview of the Next Chapter

Now that you have an understanding of why you do the things you do, you can think more clearly about how to influence and change your behavior, the theme of the next and final chapter.

Further Reading

Discussions concerning the analysis of oneself and others include:

- Richard E. Nisbett and Timothy DeCamp Wilson, "Telling More Than We Can Know: Verbal Reports on Mental Processes," *Psychological Review* 84, no 3 (May 1977): 231-259.
- Timothy D. Wilson, *Strangers to Ourselves: Discovering the Adaptive Unconscious* (Cambridge, MA: Harvard University Press, 2002).

A scholarly text on how to test for a controlled variable is:

- Richard S. Markin, *Doing Research on Purpose: A Control Theory Approach to Experimental Psychology* (St. Louis: new view, 2014).

Two publications in harmony with the approach just presented but that focus on therapists analyzing and dealing with psychological problems of their clients are:

- Lucy Johnstone and Rudi Dallos, eds., *Formulation in Psychology and Psychotherapy: Making Sense of People's Problems,* 2nd ed. (London: Routledge, 2014).
- Lucy Johnstone, Stuart Whomsley, Samantha Cole, and Nick Oliver, *Good Practice Guidelines on the Use of Psychological Formulation* (London: The British Psychological Society, 2011). Available online.

Endnotes

1. Given my dissatisfaction with the present state of psychology, I set out to develop a more unified, up-to-date theory or model of human behavior. While researching what affects human behavior and the huge number of theories and models that have been developed concerning what we do, I found that systems theory and Perceptual Control Theory expressed what I was trying to achieve—mainly for the reasons explained in Chapter 9. In particular, when I discovered PCT, it became clear that I had found a unifying framework and model that met what I was striving for, and that I could do no better than to share its perspectives with others.

2. Highlight 10.1 is a simplification, aimed at focusing attention on the main functions of perceptual control systems. It shows only one control loop. In reality, many control loops are operating at any given time to achieve desired perceptions—as mentioned in Chapter 3 and indicated by Highlight 3.7.

3. These mechanisms and factors can also be called "moderators" since they often moderate the actions of the control systems of which they are a part.

4. This reference is at what Powers calls the "program level" of the control hierarchy.

5. Klein in "An Integrated Control Theory Model," touches on some of these factors, especially those between the comparator and effector output.

6. As described in Chapter 9, PCT being the metatheory that it is, the descriptive terms used that are associated with the Person/Individual Level also seem related to the PCT framework at the Sub-Personal Level and can help readers think about possible influences on operations of PCT loop components that might not otherwise be considered.

7. A variable is "Anything that is subject to variation" according to Coleman, *A Dictionary of Psychology*, 800.

8. As with any analysis of a system, this list could be extended almost indefinitely. Therefore, only somewhat important effects are noted here.

9. Atkins, *Atkins for Life*, 4-5, 110.

10. *Obesity and Overweight*, Fact Sheet 311 (World Health Organization, updated January 2015), accessed February 3, 2015 from http://www.who.int/mediacentre/factsheets/fs311/en/.

11. I have mentioned only what seem to be more important influences since, as with any system, this analysis could be extended indefinitely.

12. Dreyfus, *Skillful Coping*, 28-29.

13 VandenBos, *APA Dictionary of Psychology*, 2nd ed., 1063.

14. Wilson and Dunn suggest this as one way of increasing self-knowledge in their article "Self-Knowledge." The approach seems worth using with others.

15. Nisbett and Wilson point this out in their classic article, "Telling More Than We Can Know."

CHAPTER 11

LESSONS LEARNED: HOW TO CHANGE YOUR BEHAVIOR

The Big Picture

The previous chapters introduced you to major influences on your behavior. You should now know that you are an autopoietic being whose control systems help you survive. You have seen that various factors affect what you do, including how you perceive things, your goals and other references, what you have learned to do, the state of your body, the influence of your environment, and interactions that occur between you and your environment.

In this chapter, you will learn how to change your behavior. You will also find out how to reduce the influence that others have upon what you do. Effective techniques include setting goals for yourself, monitoring what you do, ensuring that you can do what you need to do, and changing your environment. Resources that you may use to help you along include your family, friends, self-help groups, and even, at times, a little extra glucose in your system.

Summary of What You Have Learned So Far

The previous ten chapters have introduced you to important ideas about you and your behavior. Let's remind ourselves of that knowledge so you can use it to better change and control what you do.

Chapter 1 introduced you to the idea that you are an autopoietic being who is a very good survivor. That is, you do things that help you survive, avoid doing things that may result in your death, and have successfully done so up until now because you are organized and structured to survive. Your body's systems help you survive. Of course, your environment is also important to your survival. You need food, oxygen, and the help of others to live, but the bottom line is this: *you* *are naturally structured and organized to sense, learn, and deal with your environment in ways that help you to survive.*

Chapter 2 furthered the discussion of your body's organization. You saw that you have a number of automatic systems that help you survive—systems that maintain your body at required levels of oxygen, temperature, and energy—a process called homeostasis. Your body also has a number of automatic systems that help you avoid danger, such as processes that produce a natural fear of heights (or at least did so when you were a baby), an immune system that attacks germs that may harm you, and a "fight or flight" response that prepares you for quick action when needed. You are also structured to sense and perceive your environment, to detect pleasant and unpleasant sensations, and to learn. These and other mechanisms help you to survive and enjoy a better quality of life. Besides the idea that you have a number of automatic homeostatic systems that help you to survive, an especially important point of the chapter is that *when you do something, you do what you do in order to change or maintain your perceptions.*

Chapter 3 focused on your nervous system and how it affects what you do. In addition to your senses and perception, you were introduced to the idea of "references" and how your body does things so that your perceptions match important references that you have, including your goals and critical physiological conditions needed to survive. Perceptual Control Theory (PCT), as developed by William T. Powers, was introduced as a basic framework for understanding why you do the things you do. Related factors including your feelings, the effects of hormones, and your body's physiological condition were also discussed as influencing your behavior. Perhaps the most important point of the chapter, though, is the PCT idea that *we do things so that our perceptions match our references as much as possible.*

Chapter 4 discussed how you became the way you are. Although your genes have played a part, other major influences include interactions with your environment during your development as an embryo in your mother's womb, as an infant and child, and now as an adult. An especially important factor, however, has been learning. Your learning has occurred in various ways: through trial and error, by observing and imitating others, instruction by others, self-instruction, and simply learning by doing. Although some of this learning occurred by associating events and conditions observed, a great deal happened because it helped you to better achieve your goals and other references. In addition, your environment has played an important part in affecting what you have learned to do and not to do. "Social control" techniques, including the approval and disapproval of others, teaching and training, job descriptions, laws, regulations, and other methods reflecting the references of other people, have also influenced your learning and resulting behavior. In short, *your environment and related learning have played a great part in affecting what you do.*

Chapter 5 dealt with how regular, routine, and automatic your behavior usually is. Your personality, skills, roles, and words like norms and culture all reflect the regularity and routineness of what you do. You also learned how behavior settings such as stores, churches, libraries, schools, and workplaces are where certain expected and relatively routine behavior occurs. When what you are doing, though, does not produce the results you want, you may do something different. A key question concerned why your behavior is so regular. The basic reason is that *your regular and routine behaviors occur because they have been effective in producing perceptions that you desire—perceptions that indicate your references have either been achieved or are being achieved.*

Chapter 6 shifted our attention towards considering how you and your external environment affect one another. The idea that your behavior is a function of both you and your environment was presented, although this fact may have been evident from our earlier discussions. Influences of your natural, manmade, and social environments were discussed. So were gene, neuron, and environmental interactions; priming effects, that are ubiquitous; and how learning, your routine behavior, your body's development, and the way you feel, including your core affect, are affected by your immediate environment. A key point is that *you and your environment interact and influence one another.*

Chapter 7 returned us to thinking about perception and how you perceive your environment. You were reminded that behavior is the process by which you act on the world to control perceptions that are important to you. Your perceived environment was highlighted as essential to your survival, with your perceptions indicating available resources, opportunities for action, constraints on your behavior, and how you are doing. Perception of your environment also enables you to learn. How others try to affect and manipulate your perceptions and resulting behavior was also touched upon. A key point of this chapter is *how you perceive your environment affects what you do.*

Chapter 8 focused on the broader environment and how it affects you, even though you may not directly perceive and experience that more distant environment. Effects include the influence that millions of people you have never met have had on your life. These include impacts on you by organizations such as governments including laws passed, policies established, roads and buildings constructed, and schools established and operated; businesses, corporations and other employers; technology and science; economic conditions; and the mass media. Even Mother Nature acting faraway has indirect effects upon the environment that you experience. A main point of the chapter is that *many things both near and far affect what you experience and do.*

Chapter 9 brought us back to a more detailed examination of theories and models, and some of their characteristics and limitations. Since human behavior is so complex, the chapter emphasized that in order to understand your behavior, you need to take a holistic view of yourself, your environment, and how you interact with your environment—as this book has attempted to do. Systems theory was focused upon as a way of helping keep our eyes on "the big picture" and better understanding why we do the things we do. The chapter pointed out that, in comparison to many other theories and models of human behavior, Perceptual Control Theory (PCT) does quite well as a systems oriented, recursive, comprehensive metatheory and framework for behavior explanation that is closely linked to the biology of how our bodies operate—apparently being a revolution in scientific thought. In short, key ideas are that *systems theory and Perceptual Control Theory are helpful tools for understanding your behavior and the behavior of others.*

Chapter 10 indicated an approach based on PCT that you can use to analyze your behavior in order to better understand why you do the things you do. You considered components of the control loop including internal and environmental influences on your control loop functions. These components and other influences include your perceptions of what you are controlling, references for those perceptions, errors that stimulate your behavior, disturbances to variables you are controlling, and factors such as priming, hormones, and other influences described generally by terms such as learning, memory, expectations, framing, and reflective thoughts. In short, the chapter indicates that *systematic analysis can help you identify important influences on your behavior and resulting perceptions.*

Highlight 11.1 summarizes these main points as well as their implications for behavior change. In short, to change your behavior, you may need to do one or more of the following:

- change your perceptions
- change your references
- change your ability to do things, such as what you know, your skills, stamina, and other aspects of your body's structure and organization
- change your environment
- obtain feedback about what you do
- analyze why you do the things you do.[1]

Highlight 11.1	Lessons Learned and Implications for Behavior Change
Lesson Learned	**Implications**
1. You are naturally structured and organized to sense, learn, and deal with your environment in ways that help you survive.	1. To change what you do, you must change one or both of the following: • how you are structured and organized • your environment.
2. You do what you do in order to change or maintain your perceptions.	2. Behavior change must result in perceptions that are important to you.
3. We do things so that our perceptions match our references, as much as possible.	3. Your perceptions, references such as goals, abilities, and feedback about what you do may need to be changed in order to change your behavior.
4. Your environment and related learning have played a great part in affecting what you do.	4. You can learn to behave differently. Such learning changes the way your neural system is structured and organized.
5. Your regular and routine behaviors occur because they have been effective in producing perceptions that you desire—perceptions that indicate your references have either been achieved or are being achieved.	5. Changing your habits may be difficult since they usually occur automatically without conscious thought. Changing those behaviors must produce perceptions that are more or equally desirable and probably requires a change of the environment that triggers those perceptions, behavior, and habits.

6. You and your environment interact and influence one another.	6. You probably need to change your environment or interact differently with it in order to achieve the results you want.
7. How you perceive your environment affects what you do.	7. You may need to change how you perceive your environment in order to behave differently.
8. Many things both near and far affect what you experience and do.	8. You may need to change several aspects of your immediate and distant environments in order to change what you do.
9. Systems theory and Perceptual Control Theory are helpful tools for understanding your behavior and the behavior of others.	9. Consider "the whole picture" including your body's control loops, your environment, and how you interact with your environment when attempting to change your behavior.
10. Systematic analysis can help you identify important influences on your behavior and perceptions.	10. When trying to change what you do, attempt to identify important influences on the control loops that affect the behavior you wish to change.

A General Approach for Changing Your Behavior

Some Basic Knowledge

With the previous chapters in mind, let's move on to ways of changing, controlling, or otherwise influencing your behavior. We'll first consider some practical guidelines gleaned from what has often worked in the past for other people.

First, you should recognize that *regulating your behavior or "self-regulation" is a skill that can be learned,* and that the more you practice that skill, the better you will become. Such learning results in the reorganization of your nervous system and how you interact with your environment.[2]

Second, you should realize that *whether you change your behavior or not is up to you.* No one else can do it for you. Others can help or hinder you, but what you do is ultimately in your own hands unless either overwhelming force is used to make you do something or if what you do is biologically determined, such as the need to eventually sleep.

Third, *what you need to do will depend on the kind of behavior you wish to change.* Do you want to:

- **Stop doing something you are now doing,** such as a habit like smoking, drinking alcohol, or gambling? To do so, you will need to develop new habits, change your environment, and make plans such as how to deal with temptations, relapses, and difficult situations you may face.
- **Improve the way you do something now**, such as improving your work performance, social skills, or sports skills? This will involve learning to do things better, including lots of practice with feedback, and changing your environment to support your improved behavior.
- **Start doing something new**, such as learn a new skill or obtain a new job? If so, your focus will be on activities that will help you learn that new behavior or achieve that new outcome. Especially for new skills, you will usually need to practice the behavior involved. You may also need to change your environment to support what you wish to do, such as providing yourself with useful resources like a teacher, instructional materials, equipment, tools, and time.
- **Resist influences from others** that affect what you do? You will be required, in part, to change how you perceive things by becoming more aware of how and when others influence you.

Finally, you should realize that *establishing or changing behaviors that are not habits is often simpler than changing habitual behavior*. Changing behavior that is not a habit may simply involve setting a behavioral goal, obtaining information or otherwise educating yourself so that you know what to do and have the confidence to do it, and then behaving accordingly. Changing your habits, though, will usually require more thought, effort, and perseverance, because your habits have been learned and performed over a long period of time and are done automatically. They are ingrained. Changing habitual behavior usually involves doing two major things: (1) stopping the old undesirable habit that you are trying to change, and (2) learning new, more desirable, habitual ways of behaving.

Stopping old habits and establishing new habits requires commitment and lots of practice over a long time in order to be successful. Changing habitual behavior may be harder than you think! If you really want to change what you habitually do, don't be too flippant and casual about what is involved because then you will probably fail.

Unless what you want to do is very simple, pay close attention to the ideas in the following steps. And make sure you use them! As indicated, changing what you do can require commitment, effort, and persistence, especially if you want to change old habits and create new ones. Such behavior change is not a one-time event. The quotes in Highlight 11.2 are worth remembering.

Highlight 11.2 | **A Few Noteworthy Quotes About Behavior Change**

"People do not fail at self-modification because the techniques don't work; they fail because they don't use the techniques."

"To increase your chances of success,
Use the techniques,
Use as many as you can, and
Use them long enough to have an effect."

(Watson and Tharp, *Self-Directed Behavior*, 9th ed., 22)

"Change isn't an event; it's a process."

(Heath and Heath, *SWITCH*, 253)

Step 1: Decide to Change and Mentally Commit to Doing So

- Consider why you are dissatisfied.
- Clarify to yourself what you want to do differently.
- Commit yourself to doing what is required.

The beginning of change is natural and simple. Just think about why you are dissatisfied and want to change what you are doing. What is the problem or reason why you want to change your behavior? In technical terms, why do you have an error signal that is causing you to be dissatisfied and to consider changing what you do? Think about what your problem is and try to make clear to yourself why you wish to behave differently.

Then, decide whether the behavior that you wish to change or the problem you wish to solve is important enough for you to make the effort needed to change. If what you wish to do is important, you must commit yourself to doing what is required. It takes more than a wish to do things differently. You need to take action; and, to change more than a simple behavior, you need to be committed to doing so. Simple wishes are not enough.

Although you do not have to make a 100 percent commitment to change during this step, such commitment is needed later on if you are to be successful over the long term. If you are wishy-washy about changing or not sure, you probably will not succeed. When it comes to changing behaviors such as quitting smoking, doing things to lose weight, and increasing physical activity, successful achievers consistently report that they were committed to doing so. For example, one successful person stated, "You have to want it for yourself."[3]

The sooner you make up your mind to change, the better! Otherwise you will never complete the following steps. Doing the first steps to be described, such as clarifying your goal and planning what to do, is an indication of some commitment. But to actually complete the process, to successfully change your behavior over the long term, to overcome difficulties or relapses that you may have, you will need to be 100 percent committed to what you are doing—at least by the end of Step 5 when you make a plan to achieve your goal.

Sometimes, thinking of a higher value or principle can help you make that commitment and deal with other conflicting goals you may have. For example, in my own case, I wanted to stop smoking cigarettes. The birth of my first son helped me to do that. I realized that it wasn't fair to my wife and son for me to leave them a widow and orphan by dying prematurely from my smoking. I felt an obligation to look after them and be with them as long as possible. And so I committed myself to stop smoking and was successful. Perhaps appealing to a higher reference such as a value or principle may help you, too—be it better health, a longer life, self-respect, social respect, a sense of accomplishment, having a happier life, avoiding trouble, religious salvation, the Golden Rule, or some other reference important to you.[4]

Another way of helping yourself become committed to changing what you do is to think about the advantages of changing compared to the disadvantages of not changing. You might list them on a piece of paper. To become really committed to changing, you need to be convinced that the advantages outweigh the disadvantages. To help you decide, you might ask yourself:

- "Why do I want to make this change?"
- "What are the three best reasons to do it?" and,
- "How important is it for me to make the change, and why?"[5]

In other words, think about why you want to change. Become committed to achieving the change that you want. ***Make up your mind to do it!***

Step 2: Decide On a General Goal to Achieve.

Once you have an idea why you are dissatisfied and want to change, the next step is to clarify your goal. Do you want to do something such as lose weight? Stop smoking? Learn something new? Stop procrastinating? Stop doing something else? At this point just a general idea of your goal is OK. Nothing specific is needed.

Highlight 11.3 Some Common Goals Involving Behavior Change

achieve financial security
adopt a low-salt diet
anger control
be a better parent
become more assertive
change cholesterol levels
control nervous habits (such as nail biting)
date more
deal with post-traumatic stress disorder
eat less junk food
eliminate teeth grinding
exercise more/increase physical activity
find a job
handle anxiety in social situations
improve one's marriage
improve sport performance
improve test-taking skills
lose weight
manage diabetes
manage medication regimen
overcome alcohol addiction
overcome anxiety
overcome depression
overcome drug addiction
reduce blood pressure
reduce conflicts with co-workers
reduce fear (such as of speaking in public)
reduce panic attacks
reduce procrastination
reduce stress levels
sleep better/reduce insomnia
stop gambling
stop smoking
study more effectively

Major Sources: Watson and Tharp, *Self-Directed Behavior*, 9th ed.; Leutzinger and Harris, *Why and How People Change*; Gatz, Smyer, and DiGilio, "Psychology's Contribution."

To help you along, Highlight 11.3 contains a list of common goals. These examples are of two general kinds: (a) *behaviors* that many people wish to change, such as stop smoking or eat less junk food, and (b) desirable *outcomes* they wish to achieve that requires them to do things differently, such as lose weight or change cholesterol levels. For now, either kind of goal is OK. But later on, if you have an outcome goal, you'll need to eventually clarify what you need to do to achieve that goal. Even if your goal is not a specific behavior, reaching the goal will require changing some of your behavior.[6] For now, though, just decide on your general goal.

Step 3: Gather Information

- Obtain information about yourself, situations, and people, related to your goal.
- Obtain information about how others have achieved similar goals.

The idea here is to gather information that will help you achieve your goal. This information is of two types: (a) information about yourself and the situations you encounter, related to your goal, and (b) information about what others have done to successfully achieve similar goals or change similar behaviors. Gathering such information will help you better plan what to do[7] and increase your chances of success.[8]

The first type of information requires that you *think about and observe yourself with regard to your goal.* For example, if you want to stop an undesirable behavior, analyze the behavior using the approach described in Chapter 10 and the guide shown in Appendix 2.

Among other things, when dealing with an undesirable behavior and considering your environment, observe what triggers or seems to initiate the behavior you are concerned about. Become aware of what people and situations, if any, make the behavior more likely to occur and what situations make it less likely to occur. For example, if junk food is lying around, do you eat it when you see it? Do you exercise more when someone you are close to wants to do so also? In addition, what internal thinking and moods, if any, trigger the behavior? For instance, does feeling sad, tired, or angry cause you to do it? Do you feel better or worse after the behavior occurs? Why? Are you rewarded or punished in some way?

Watson and Tharp, who have written about behavior change, suggest that you make a list of concrete examples of the problem, and even keep a diary of when things happen, details of your problem, and your thoughts and feelings, rather than just vaguely thinking about what happens. They state, "Self-knowledge is the key to successful self-change" and that "in order to change yourself, you must know what you are doing."[9] For example, if you have a weight problem, you may notice and record that you keep a ready supply of fattening foods in the house, you eat to avoid wasting food, you put a lot of food on your plate, you eat very rapidly, you eat when you are tired or emotionally upset, you eat a lot when you go to restaurants, you eat many times during the day, you rarely weigh yourself and do not exercise enough, you often skip breakfast, and you sometimes starve yourself and then overeat.[10] Observing and becoming aware of such factors will help you to deal with them later by providing a focus for your planning and action. For example, you may need to remove or avoid the environmental triggers that set off your undesirable behavior, or arrange new triggers to initiate more desirable behavior that you wish to develop.

Regarding how others have dealt with similar behavior change, try to benefit from the experiences of others and not waste your time doing things that will not help you. You should, for example, try to find out what others have written—online, in articles, and in books—about achieving a goal such as yours. You might also talk with and perhaps observe other people who are dealing with or have successfully achieved the kind of results you are interested in. "Highlight 11.16: Examples of Self-Help Resources," discussed later in this chapter, may help to point you toward useful sources of information.

Step 4: Clarify Your Goal

- Your goal should be specific, measureable, achievable, relevant, and time bound.

Self-regulation and control involves guiding yourself toward achieving or maintaining a goal. Sometimes, people have found that it is more workable to set a modest goal at first such as to lose 1 pound a week, or to follow the Alcoholic Anonymous mantra, "One day at a time," rather than to set goals that are too large such as to lose 80 pounds, or stop drinking alcohol forever.[11]

Clarifying your goal is necessary, both from a practical as well as from a theoretical point of view. Research has shown that doing so is one of the most important things you can do to change your behavior.[12]

The clarified goal that you now set should be SMART:

- **S**pecific: not too general or vague.
- ***M***easureable: capable of being able to determine if you are making progress.
- **A**chievable: something that you believe you can accomplish, is not too ideal to be attained, and you are willing to work towards.

- ***R****elevant*: something that you value, want to do, and is truly important to you.
- ***T****ime-bound*: such as when you will do things and/or a target date for completion.

For example, a vague goal would be to "lose a lot of weight" or to "improve how fit I am." A SMART goal would be to "lose one pound of weight a week for 20 weeks," or if you wanted to improve your fitness, "I will work out at the local gym for at least 30 minutes three times a week at 7:00 pm on Monday and Wednesday and at 10 am on Saturday."[13]

In addition, try to ensure that your goal does not conflict with other important references that you have. Otherwise, you will increase your risk of failure. Baumeister and Tierney, who have written a book on willpower, also suggest that you focus on only one self-improvement project at a time and then stick to it, rather than trying to accomplish a list of different resolutions.[14] So, try to achieve only one major change goal at a time. After you are successful with one goal, move on to another, if you want to.

Highlight 11.4 Goals: Some Notable Quotes

"A goal without a plan is just a wish!"

(Antoine de Saint-Exupery, French writer, 1900-1944)

"Setting measurable goals provides a 'yardstick' against which to chart progress and affirms your commitment to achievement."

"Take it one day at a time — Seeing a change as something you are going to do for the **rest of your life** is overwhelming. Seeing the change as a series of day-by-day steps makes it easier to accept and comprehend."

(Leutzinger and Harris, *Why and How People Change*, 8)

Step 5: Make a Plan to Achieve Your Goal.

In this step you select and organize the methods that will help you achieve your goal. Do the following when planning the actions you will carry out:

- Specify the behaviors involved in achieving your goal.
- Decide how to monitor your progress.
- Decide what environmental changes you need to make.

- Consider how you will avoid undesirable triggers and temptations.
- Plan to deal with threatening and opportune situations you may face.
- Prepare to deal with slips, errors, and lapses.
- Consider using support persons, groups, technology, and other resources.
- Ensure that you have the knowledge, skills, commitment, and resources needed to achieve your goal.
- Reward yourself for success along the way.
- Write down your plan.

For more difficult behavior change, you will need to use most of these techniques. For a simple change, only a few may be needed. Let's look at each in more detail. You can then decide what mixture will best help you to achieve your goal.

Specify the Behaviors Involved

By this time you should be aware whether your goal is a behavior or the outcome of behavior. For example, losing weight is an outcome. But it is not a behavior. Achieving such an outcome is a result of changed behavior such as eating less food, drinking fewer sugary sodas, and exercising more.

The behaviors that you specify will depend on the information you gather in Step 3, the goal you set during Step 4, and what you decide to do after reading the possibilities described in this step. For example, if your goal is to lose one pound of weight a week, depending on the information you gather and thinking about what follows, you may decide to do things such as:

- Weigh yourself every day.
- Give away junk food and other fattening food that you have in the house and do not buy more.
- Let your family, friends, and co-workers know that you are trying to lose weight and seek their support.
- Do not take second helpings of food.
- Use smaller plates and bowls when you eat at home.
- Eat more slowly.
- Drink two or three liters of water and/or other fluids a day.

- Do something besides eating food when you are tired or upset.
- Decide beforehand what you'll order before you go to restaurants.
- Walk for 30 minutes every day.
- Eat breakfast every day.
- Limit snacks to small healthy portions at mid-morning, mid-afternoon, and before you go to bed.

If doing all this seems overwhelming, phase in what you do, one behavior at a time. For example, the first week you can weigh yourself every day. The second week, continue to do that and also give away junk food and other fattening food you have in your house—and don't buy more. The third week, continue doing those things and, in addition, let your family, friends, and co-workers know that you are trying to lose weight and ask them for their support. As one change agent, John C. Norcross, put it, "Small steps together equal a giant leap. Start small and then incrementally increase your sub-goal activity."[15]

Decide How to Monitor Your Progress

Besides setting goals, monitoring progress toward them is one of the most important things people can do to change their behavior.[16] The more often one monitors and obtains feedback the better. Set up a way to monitor your progress, such as count each time you do or don't do something, measure important variables, or keep a diary. Then compare the results with the goal or sub-goal to which they are relevant. The monitoring techniques you use will depend on the goal and sub-goals you wish to achieve. For example, if you want to change a habit, vigilant monitoring of unwanted behavior (thinking "don't do it" and watching carefully for slipups) can help to suppress their occurrence and enable the development of new, more desired behavior.[17]

Write down the results, especially those concerning achievement of your major goal in order to see the progress you are making. Making a chart or graph that others can see is a good idea. Doing so can help to hold you accountable and increase your commitment toward reaching your goal. Start by writing down your present baseline measurement or measurements. For example, if you wish to lose weight, write down your current weight. Once you begin to take action aimed at implementing your plan, measure your weight regularly. If you wish to stop smoking, keep a record of how many cigarettes you now smoke each day. Then, continue record-keeping once you begin your plan to reduce or stop smoking.

In addition to monitoring your progress toward achieving your goal, you should also monitor implementation of your plan. That is, make a note to yourself to regularly check on whether you are implementing the different parts of your plan as intended. Doing so will help ensure that you do not forget to do important things. You may also obtain information about how effective those things are and whether changes are needed to make your approach as effective as possible.

Highlight 11.5 Examples of Monitoring Techniques

Count the number of times you do something:

- Record a tally mark each time, on an index card, in a small notebook, or next to items on a list.
- Click a button on a counter each time a relevant behavior or thought occurs.
- Use a pedometer when walking to measure the number of steps you take.

Record the amount of time you spend doing something:

- Keep a record of your starting and ending times.
- Use a stopwatch.

Measure important variables:

- Your weight for weight control
- Blood glucose self-monitoring; routine urine testing for diabetes control
- Blood pressure for hypertension control
- Forkfuls of food consumed and speed of consumption for weight control, possibly using the Hapifork[a]

Keep a record of how much you do different things:

- Record specific activities that you do each time you do it, such as exercise activities or the number of pages read.
- Keep a diary of your thoughts and feelings.
- Use a timer or other electronic device to remind you to record what you are doing at the instant it makes a sound.
- Keep a food diary of what you eat and record each time you eat something and your intake of calories or carbohydrates.

- Keep a record of people, objects, or events that trigger an undesirable or desirable target behavior.
- Take photos of yourself.
- Make audio or video recordings of your behavior and listen to or look at the recordings.
- Use self-monitoring software and web site tools.
- Use a wearable computing device, such as a Fitbit.[a]

Other monitoring techniques:

- Be mindful of what you are doing.
- Ask others to give you feedback.

[a] This is described in a later section entitled "Consider Using Technology."

Monitoring your progress and obtaining feedback will help you (a) continue doing things that are assisting you to progress towards your goal, and (b) change what you are doing if you are not making enough progress. Without feedback, you will act blindly and probably fail in what you are trying to do. In addition, monitoring keeps your attention on your goal, making it almost impossible to forget. You may also find it highly rewarding and satisfying to see the progress you are making![18]

Decide on Environmental Changes

Changing your environment is important, especially if you are eliminating a habit or starting a new one, because your habits are closely linked to your environment. Aspects of your environment tend to trigger your habitual behaviors. And so, if you wish to stop a habit, you will probably need to deal with those triggers. Why? It goes back to Chapter 1 and your autopoietic being. The outside environment may trigger actions on your part, depending on your internal organization and structure at that moment. These actions are called habits when they occur regularly, fairly automatically, and usually without much thought. As a result, to change a habit, you need to either (a) change your control systems' structure and organization such as by developing new habits, (b) change your environment so that it no longer triggers the undesired habit, or (c) do both.

Changing the structure and organization of your nervous system takes time—often over a month or two to change a habit. Because of this, you probably need to change your environment and the triggers within it or you may relapse and revert to the habitual behavior you are trying to change. Removing or avoiding triggers for your old habit will help prevent it from occurring again.

In addition, establishing triggers or cues for a new habit will remind you to perform the new behavior. A newly established trigger will become linked to the new habit so that when that trigger occurs, you will be reminded or otherwise influenced to carry out the new behavior.

Of course, other things besides environmental triggers affect what you do, including how easy or hard it is to do something. And so, also consider (a) making it harder to perform the undesirable behaviors you are trying to reduce or eliminate, and (b) making it easier to perform more desirable behaviors you want to establish or increase. To do this, you usually need to change your environment.

Highlight 11.6 gives an overview of some techniques you can use to eliminate old triggers, establish new triggers, make it harder to perform behaviors you want to reduce or eliminate, and make it easier to perform new behaviors you want to establish.

Highlight 11.6 Ways to Change Your Environment

To increase desirable behavior by changing environmental triggers and making it easier to perform desired behavior:

- Arrange helpful cues, signs, reminders, prompts, and other nudges.
- Remove obstacles, such as too many intrusions, lack of time, lack of information and skills.
- Make it easier to perform the behavior, such as simplifying procedures, obtaining help, and/or using appropriate technologies.
- Provide yourself with needed or useful resources like equipment, tools, and time.
- Use support people or groups.
- Associate with people who encourage or perform the behavior you desire.
- Obtain feedback by using technology or other sources.
- Change your location; move to a new place that is more supportive of your desired behavior.

To decrease undesirable behavior by changing environmental triggers and making it harder or impossible to perform the undesirable behavior:

- Avoid people, things, places, and events that trigger the undesirable behavior or otherwise tempt you to do it.
- Remove objects from your immediate environment that are needed to perform the behavior.
- Obtain feedback about what you are doing in order to make yourself more aware of your undesirable behavior and its outcomes.
- Change your location; perhaps move to a new place.

The following sections go into detail about some of these techniques and the practicalities of changing your behavior. Do the best you can, but recognize that you may not be able to change everything that should be changed in your environment. In some cases, you may not have adequate power to change conditions that others have created, especially when an imbalance of power exists. If you are an employee, a student, an aged person, or a young person with parents, you may be limited in the things you can do.[19]

Consider How to Avoid Undesirable Triggers and Temptations

Think how you will remove or avoid undesirable triggers, conditions, and temptations that lead to behaviors you are trying to reduce or eliminate. For example, if you wish to lose weight, you might remove junk food from your house and avoid going to buffet restaurants. If you wish to stop drinking alcohol, do not keep alcohol where you live and do not go to bars or even near them, if possible.

Many triggers are external, such as environmental cues, friends, requests, offers, seeing what other people are doing, the time of day, contexts and locations, and situations. Some are internal, such as your moods; feelings of hunger, stress, and anger; thoughts, such as about a traumatic event; and how you perceive an object or situation. Some are prior behaviors in a sequence or program of behavior, such as eating a meal that triggers the habit of having a cup of coffee or a cigarette afterwards. In fact almost anything can become associated with a learned behavior and serve as a trigger for that behavior. Besides directly triggering your habitual behavior, some objects and events may also prime your neural networks and make it more likely that you will do certain things.

The reason for dealing with triggers, temptations, and unwanted priming is to avoid activating your undesirable behavior. For example, seeing junk food in your home may trigger or prime you to eat it—something you may not want to do if you are trying to lose weight. Going to the mall or reading advertisements for sales may prime you to shop—another undesirable behavior if you are trying to reduce a compulsive buying habit that is putting you in debt. Perceiving a second helping of food as being desirable compared to seeing it as adding to your weight or perceiving a beer as something that will help you to relax compared to something to be avoided if you are an alcoholic trying to stop drinking alcohol may also trigger behaviors you are trying to change. In such cases, your perceptions may need to be changed. However, you should also avoid or eliminate the triggers and temptations that set off those undesirable perceptions and behaviors.

If your trigger is an internal state such as stress, you might try to reduce or eliminate it by learning techniques for reducing stress levels such as meditation that David Gamow describes in his book *Freedom from Stress: How to Take Control of Your Life*.[20] You might also try to avoid situations that cause stress. Or, according to Robert M. Sapolsky, you might seek social support, exercise, or practice meditative techniques to alter your perception of the stressor.[21] Similarly, if an important trigger is another internal state, such as anger, depression, or hunger, you might (a) make control of that state a sub-goal to be achieved, (b) obtain information about how to control or deal with that state by reading relevant materials or talking with knowledgeable people, (c) plan what to do, and (d) take action to achieve that sub-goal.

Alternatively, you may arrange or do things to encourage priming and triggering effects in order to activate more desirable behavior that you hope to establish. Writing a reminder on the calendar or setting an alarm on your cell phone may remind you to go to the gym or do another important activity. Similarly, associating with others who value or are doing what you want to do, may trigger values, norms, and related action. Behaviors and perceptions in a sequence of events can also serve as useful triggers. For example, as discussed in the next section, you might develop an "if-then" plan so that "if" something occurs—the trigger—you will "then" do a desirable behavior you have planned to do.

According to B. J. Fogg, focusing on triggers is the simplest type of change you can do and is often all that is needed to change some behaviors.[22] So do try to (a) identify and eliminate or avoid triggers of undesirable habits and other behaviors that you wish to change, and (b) identify and establish triggers for desirable behaviors and habits that you want to occur.

Highlight 11.7	Triggers and Habits

"Without a Trigger, the target behavior will not happen."

(Fogg, *Triggers*, 1)

"… the dependence of habits on environmental cues represents an important point of vulnerability. Disrupting the environmental cues that trigger and maintain habit performance renders habits open to change… ."

(Verplaanken and Wood, "Interventions," 91)

Develop "If-Then" Plans

You are bound to be faced with an unavoidable trigger, temptation, or other situation that threatens to interfere with what you are trying to change. This is natural. However, there are ways to deal with such occasions. The best way is to plan beforehand what you will do when you face such a situation. That is, make an "if-then" plan to guide what you do.

These plans take the form that "if" you perceive a certain situation, "then" you will take a certain action. For example, when trying to control your weight, you may decide in advance that "if" you end up at a buffet, "then" you will take only one plate of modestly portioned food. Or when eating at a friend's house, "if" you are offered a second helping of food, "then" you will say, "No thank you. I took enough the first time." Similarly, in case you are trying to control your angry outbursts, you may decide that "if" you perceive you are getting upset, "then" you will "take 10 deep breathes" or "if" you have time, you will "sleep upon the matter" before taking action.

Having such if-then plans, or "implementation intentions" as the psychologist Peter Gollwitzer calls them,[23] will help you to achieve your goals since they take the guesswork out of what to do when you are in a troublesome situation. Just do what you have decided to do beforehand.

Besides helping you deal with undesirable triggers and other temptations, if-then plans can help you develop desirable habits too. For example, when you have a goal such as sleeping better, you might have an if-then plan such as "if it is later than 4:00 pm," "then I will not drink coffee, tea, or other drinks containing caffeine"; and perhaps, "if it is 9:00 pm," "then I will turn off the TV, computer, and other electronic devices and dim the lights around my house." In such cases, the "if" part of the plan acts as a trigger to set off the desired "then" behavior you want to perform. When this occurs enough times, your desired behavior will become a habit!

Such if-then plans have been shown to be effective in promoting goal attainment when they are based on a strong commitment to both achieve the underlying target goal and to implement the plan. [24]

Highlight 11.8 **If-Then Plans**

"...if-then plans (i.e., implementation intentions)....produce automatic action control by intentionally delegating the control of one's goal-directed thoughts, feelings, and behaviors to specific situational cues."

(Gollwitzer, Bayer, and McCulloh, "Control of the Unwanted," 485)

"An 'if-then' plan requires you to specify exactly what you are going to do and when and the situation you are going to do it in."

(Michie et al., *Improving Health: Changing Behavior*, 53)

"The advantage of an 'if . . . then' plan is that you don't have to rely on making a good decision while you are under stress. You have already made the decision about what you are going to do, and now you just carry it out automatically."

(Watson and Tharp, *Self-Directed Behavior*, 9th ed., 45)

Prepare to Deal with Slips, Errors, and Lapses

If you are trying to change anything but a simple behavior, you may have setbacks and make mistakes. Expect them. As David Watson and Roland Tharp point out, "When old unwanted behaviors have been automatic for a long time, everyone slips back into them at unguarded moments. After all, you are trying to change habits that you may have followed all your life."[25]

The important question is "How should I respond to my mistakes?" The answer is that your mistakes are only a disaster if they cause you to stop trying to change.[26] *The way to respond is to use them as learning opportunities.* Think about why the relapse or mistake occurred, learn from the experience, and try to avoid doing it again. For example, perhaps you need to avoid the risky situation that led to your relapse. Maybe you need to make an if-then plan to deal with similar situations that you can't avoid. Perhaps you need more practice. Possibly a trigger or barrier that you had not thought of before needs to be dealt with. Maybe you are not self-monitoring enough or your goal is not realistic and needs to be changed.

One good if-then plan that you should have is that "if" I have a lapse, "then" I will try to learn from what happened and continue striving toward my goal. To paraphrase an idea contained in a book by Christopher Hadnagy recommended for further reading at the end of this chapter: Behavior change is like mastering the art of cooking. By mixing the right ingredients in the right quantity, you can have a meal that is full of flavor and excitement. The first time you try to cook a meal, it might have too much salt or it might lack flavor, but you don't immediately give up cooking—you keep trying until you get it right. The same goes for behavior change. [27]

Highlight 11.9 **Other Notable Quotes About Slips and Mistakes**

"Behavior change is... a process of slipping, learning from the mistake, and trying again."

"With the help of a good plan for coping with slips,...you can make a mistake without derailing your entire goal."

(Norcross, *Changeology*, 152,154)

"Mistakes are simply feedback about the need for more practice."

(Watson and Tharp, *Self-Directed Behavior*, 9th ed., 41)

Consider Using Support Persons and Groups

Think of people who are likely to be helpful and support what you are trying to do. They may include those around you such as family and friends and others such as a campus counseling service or self-help group. For example, you might tell someone close to you about your goal, like a spouse, partner, or close friend. You may even announce your intention to a group of people that you know, such as friends or co-workers. Doing so may help you keep working toward that goal, rather than facing the embarrassment of quitting. Members of the group may also encourage you to keep trying and help energize your efforts.

Your support persons can be of active help, too, such as looking after your child if you need time to do a goal-related activity. If the person is knowledgeable, you can talk with him or her to obtain advice, guidance, and comfort when you are facing difficulties or have questions. Some organizations such as Alcoholics Anonymous (AA) recognize the value of knowledgeable support. AA encourages its members to find an experienced fellow alcoholic, called a sponsor, to help them understand and follow the program. You might also have a support person or two review your plans to help ensure they are workable.

You are encouraged to consider using the following "stepped approach."

1. First, try to change or control your behavior yourself, with some support and assistance from others such as family and friends as needed. If that is not sufficient, go to Step 2.
2. Use self-help groups and/or materials. If that does not seem to do the job, go to Step 3.
3. Obtain professional help.

Activities 2 and 3 above are discussed later in this chapter. For now, focus on the first step to change or control your behavior with some support and help from others.

Highlight 11.10 Social Support

"Seek support from others — Family members, co-workers, support groups, workplace health promotion programs, and other means.... If you have the support readily available to you, use it! If you do not, find ways to add support to your life."

(Leutzinger and Harris, *Why and How People Change*, 9)

Consider Using Technology

Technologies exist that may help you achieve your goals. These include something as simple as an alarm clock that can remind you to do things; counters such as pedometers to record the number of steps you have taken; wearable electronic devices to keep track of your energy expenditure; nicotine chewing gum and transdermal patches to help you stop smoking; smart phone apps that provide guidance or can be used to keep records; computer assisted instruction; internet-delivered interventions and monitoring programs; and other online information, resources and support, such as the stickK approach for increasing and maintaining your commitment (soon to be discussed) and www.43things.com, that you might look at. For example:

- **Clocky** is an alarm clock that runs away and hides if you don't get up when it goes off (see http://clockway.com for details).

- **Fitbit One™,** a clip-on device that tracks steps taken, distance traveled, calories burned, floors climbed, hours slept, and awakens you in the morning (www.Fitbit.com).[28]
- **Hapifork** measures the amount of fork servings per minute, intervals between fork servings, and how long it takes to eat your meal (http://www.hapilabs.com/products-hapifork.asp).
- **Smartphone Apps** include those for monitoring exercise, counting calories, personal finance, alcohol use, smoking, mood monitoring, reminders and organization, and relaxation (for details see http://www.changeologybook.com/resources/our-favorite-smart-phone-apps-for-self-change).
- **Online self-help resources**, available from groups and organizations indicated later in Highlight 11.16.

Depending on your goal, these or other technologies may be worth looking at for help with monitoring your behavior, triggering what you do, and providing other assistance.

Drugs and other medical procedures might also be used. Besides nicotine chewing gum and transdermal patches to help stop smoking, and bitter-tasting nail polish to help stop nail-biting, other products and techniques may be helpful, especially if your own efforts are not working. For example, disulfiram (Antabuse) induces nausea and headache when alcohol is imbibed. Dopamine-related drugs and deep brain surgical stimulation can help control Parkinson's disease symptoms. Antipsychotic drugs may help reduce symptoms of schizophrenia such as hallucinations, delusions, and disorganized or incoherent thinking. Antidepressants and mood stabilizers can help deal with depression and mood disturbances such as bipolar illness, and Ritalin can help with attention deficit hyperactivity disorder.[29] Drugs and surgery can also help you lose weight. Such approaches, however, require professional help and are perhaps best considered if your own efforts to change are not effective.

Ensure That You Have the Ability

Ensuring that you have what you need to achieve your goal is, perhaps, self-evident. But be sure you do it! If you need to learn something to successfully achieve your goal, learn that thing. If you lack the confidence and commitment to reach your goal, build it up. If you need resources such as time, information, or tools, make efforts to obtain them. Otherwise, you will fail to achieve your goal. The following sections indicate how to obtain the capability you need.

Knowledge and Skills

Hopefully, as part of Step 3, you gathered some information about what works by reading relevant materials and perhaps talking with people about how they coped with similar problems. If so, you have a good idea of what's involved in making the kind of behavior change that you want to make. If not, find out what works now!

Besides that knowledge, you also need to have the skills to do what is required. Needed skills may be developed in various ways, such as through self-study, enrollment in formal courses, through on-the-job training, by participating in a self-help group, or by obtaining a coach, counselor, or therapist to help you progress. Besides physical and academic skills, you may also need skills in often overlooked areas such as an ability to perceive things that you need to perceive such as internal and external triggers of behavior that you want to avoid, triggers of behavior that you want to develop, and awareness of undesirable behaviors when they occur. Mindfulness and awareness training is one way to develop such skills. Such training can help you self-monitor better in order to avoid slips or to catch them as they are made.

Self-control or willpower is another ability that you will need. This ability can be built up by exercising it. That is, the more you use self-control over days, weeks, and months, the better you will be able to control what you do.[30] That seems to be why mindfulness training and regular meditation over time helps to develop self-control. They are like a muscle that becomes stronger the more you use it.

Unfortunately, the opposite is true over a short period of time, such as a day. That is, if you use self-control a lot earlier in the day, later on during that day you may be less able to exert as much self-control over what you do.[31] For example, if you have a busy, stressful day at work, you may more easily lose your self-control at home with your spouse, partner, or children. However, loss of control can be reduced by not exhausting yourself trying to do too many things, by being more self-aware of what you are doing, by getting enough sleep, and even by having a snack or glucose or sucrose supplement such as a sweetened drink to keep up your energy supply.[32]

Confidence and Commitment

Other abilities you need are the mental confidence and commitment to do what is required—to start to take action, follow your plan, revise it when required, and make the long-term, lifestyle changes needed if the behavior change is to be a continuing part of your life. Ways to develop that confidence and commitment if you don't have it already include:

- **Do a pro-con analysis:** List the advantages and disadvantages of changing your behavior. Do not start to carry out your plan until the advantages outweigh the disadvantages!
- **Higher references:** Think of important references that you have such as maintaining the love of a special person, a commitment to family, honesty, social acceptance, a need to be respected by others, an ideology that you strongly believe in, or other important values related to achievement of your goal. Focusing on these higher references can help solidify your commitment to achieving your goal.
- **Early success:** Start with achieving small sub-goals and experience early success in what you do.
- **Be persuaded by others:** Observe, talk with, or read about others who have successfully achieved goals similar to yours. Doing so may help persuade you that you, too, can do the same. For example, a book by Norcross and others titled *Self-Help That Works* has ratings of autobiographies and films that may be helpful.
- **Associate with helpful people:** Initiate contact with like-minded buddies or models, such as a gym partner or a support group; have someone that you can call upon for help when your commitment wavers.
- **Go public:** Let others know of your goal and plans. Perhaps hang up your progress charts on a wall so others can see them. As Robert B. Cialdini points out, "Public commitments tend to be lasting commitments."[33]
- **Consider public contracts and penalties:** An example is the stickK approach to precommitment described in Highlight 11.11. Or you might formally sign a behavior change contract with someone and pay a penalty if you fail to achieve your goal.

Importantly, letting others know about your goal and plans can help to increase or solidify your commitment. Making promises and signing pledges can have the same effect. Although you might let others know what you are trying to do, the psychologist John C. Norcross advises, "Don't go public if you are not prepared; better yet, don't go into action if you are not prepared."[34]

Highlight 11.11 stickK: An Approach to Precommitment

The stickK approach is a way to help you to make a commitment and stick to that commitment. The approach involves going online and making a Commitment Contract by (1) stating a goal or mini-goal that you want to achieve, (2) stating an amount of money, if any, you will pay to a friend, foe, charity, or an organization that you detest if you do not achieve your goal, (3) designate a referee to confirm the accuracy of your reports to stickK, and (4) indicate the e-mail addresses of supporters you may choose, if any, to help cheer you on.

Although you do not have to put money on the line, research has shown that the commitment and success rate of those who do is much higher than those who do not. Also, although you can self-referee as a part of (3) above, your commitment and chances of success will be higher if you have someone else do so.

I used stickK to help me successfully complete a draft of this chapter. You too can use it to help you achieve your goals. To start, simply go online to www.stickK.com.

Resources

Specific resources may also be needed to ensure that you have the ability to perform as desired. They may include *materials* you need, such as condoms, for safe sex, if that is your goal; *time* to do what you need to do; *information*, such as feedback from others about how you are doing; *tools*, such as an alarm clock to get up on time or to remind you to do something; *authority* required, for example, at work if you want to do something new; and the *cooperation* of others. For example, if time is needed so you can go to the gym or go to a self-help group meeting, you might ask someone who supports you, such as a family member, to look after your children or to prepare a meal.

Alternatively, if you wish to decrease or eliminate an undesirable behavior, you should take away resources needed to perform the undesirable behavior. For example, if you are trying to lose weight, you should remove undesirable food from your house.

Reward Yourself

As you make progress toward your goal, consider rewarding yourself for small successes. You might even consider an especially big reward for yourself when you succeed in achieving your final goal.

Rewards do not have to be expensive. They can be as simple as taking pride in what you have done and telling yourself, "Good! I did it," each time you perform a planned behavior.[35] Or you could make performance of the behavior more interesting, as a man who needs more exercise might do by joining a Zumba class that contains mostly women.[36] You might also save up for rewards, such as setting aside $1.00 or $5.00 each time you achieve a desired behavior. Then, at the end of the week or month, spend the money on a special reward for yourself, such as a nice meal with friends, going to a special sports event, or musical show.[37]

Highlight 11.12 Examples of Rewards

Rewards that do not cost money	Rewards that cost money
Saying, "Well done," to yourself	Buying yourself a CD, DVD, or app
Inviting friends to your home	Buying new clothes
Listening to music	Going to a movie
Going for a walk	Buying sports equipment
Watching a favorite TV program	Going to a sports event
Taking a relaxing bath	Going out for a meal
Doing some gardening	Buying yourself perfume or a special tool
Asking your partner to give you a massage	Booking a holiday or weekend break
Listening to praise from friends or family	

Major Source: Michie, and others, *Improving Health*, 37.

The idea is to continue encouraging yourself to succeed by rewarding yourself when you have done well. Such rewards can help you establish desirable habits that you are seeking to instill—especially at the beginning of your change process. Ideally, however, external rewards will not be needed later on when your own internal satisfactions are sufficient for achieving a goal that you especially value.

Write Down Your Action Plan

If your plan is not simple, write it down so you will not forget what to do. Include your SMART goal, behaviors involved in reaching your sub-goals, how you will monitor your progress, environmental changes you will make, and other details appropriate to your situation. These may be how you will deal with triggers and temptations, your if-then plans if any, support persons and technology that you will use, how you will ensure you have the ability you need to proceed successfully, rewards you may give yourself for success, and a starting date.

Look at the "Planning Guide for Behavior Change" shown in Appendix 4. You may find it a useful template for your thoughts. If so, photocopy the guide. On it, write down your goal and details of your plan, and then use it as a helpful reminder of what to do.

Step 6: Carry Out Your Plan

In this step you do what you have planned. What's involved should be familiar from reading the previous step and making your plan. However, the following points are worth noting.

There is no consensus on exactly how to begin this step. Some specialists feel that you should start slowly, taking small steps at first—steps that will ensure your success from the beginning. For example, if you are planning to lose weight, the first week you might spend getting rid of junk food in your house. The second week, you might make sure that you eat breakfast every morning. And continue to gradually carry out parts of the plan you have developed. Other specialists suggest that you start by implementing as much of your plan as you can to ensure that the behavior you wish to establish actually occurs. The decision whether to have a fast or slow start is yours. Either way, *take action* and don't just sit around hoping and planning to change!

Also, remember, you do not want to lower the possibility of success by (a) trying to change several different major behaviors at one time, or (b) depleting your energy by not getting enough sleep and good nutrition. Don't take attention away from what you aim to do by trying to achieve several different change goals at once or by not having enough stamina to resist temptations and to do what your plan requires.[38]

In addition, don't forget that to achieve long-term behavior change, build habits. Doing so requires practice. So consciously practice the new behaviors you wish to establish until they become automatic. This may require 30 to 60 days or more to achieve. Do not slacken your awareness and efforts. Keep repeating and consciously practicing the behaviors you desire! Highlight 11.13 provides a suggested timeline for long-term success.

Highlight 11.13	A Timeline for Long-term Success
Step 1: Decide to change and commit yourself to doing so. Step 2: Decide on a general goal to achieve. Step 3: Gather information. Step 4: Clarify your goal. Step 5: Make a plan to achieve your goal.	7 to 21 days
Step 6: Carry out your plan. Step 7: Persevere and revise your plan as needed.	Up to 90 days or more
Step 8: Maintain your gains.	From day 90 onwards
Note: This timeline is similar to one suggested by Norcross, *Changeology*, 36.	

Step 7: Persevere and Revise Your Plan as Needed

- Expect and deal with lapses, mistakes, and failures.
- Be flexible and revise your plan as needed.
- If what you are doing is not succeeding, use a "stepped approach" and contact a self-help group or seek professional help.

Lapses and other setbacks are to be expected. Many people have them when trying to change their behavior. As Watson and Tharp point out in their book, *Self-Directed Behavior,* "You are learning a new skill, and skill development requires practice to smooth out the mistakes."[39] When they occur, treat each lapse or setback as a learning opportunity. *Try to understand why you had the lapse,* its causes, and then plan to deal with the cause if you face it again in order to prevent a future lapse or setback. Doing one of the following should be sufficient: (a) avoid the cause in the future, (b) develop an appropriate if-then plan, or (c) change your approach based on what you have learned.

The main idea is to be flexible. If something is not working, change your plan in order to do better. Plans are only guides to action. If some part of your plan doesn't work, change it! If an addition is needed, add it! For example, if you discover a high-risk trigger or situation that you hadn't considered before, try to avoid or remove that trigger or develop another if-then plan to deal with it. Or maybe you'll find that your goal needs to be revised. If so, revise it. Or perhaps you have forgotten to carry out some part of your plan. If so, try to carry it out. Or perhaps you've neglected social support that can help you, from your family, friends, or co-workers. If so, consider how they might help—perhaps with reminders, feedback, or encouragement. Then talk with them and try to obtain that support!

Highlight 11.14 **A Common Pattern of Behavior Change**

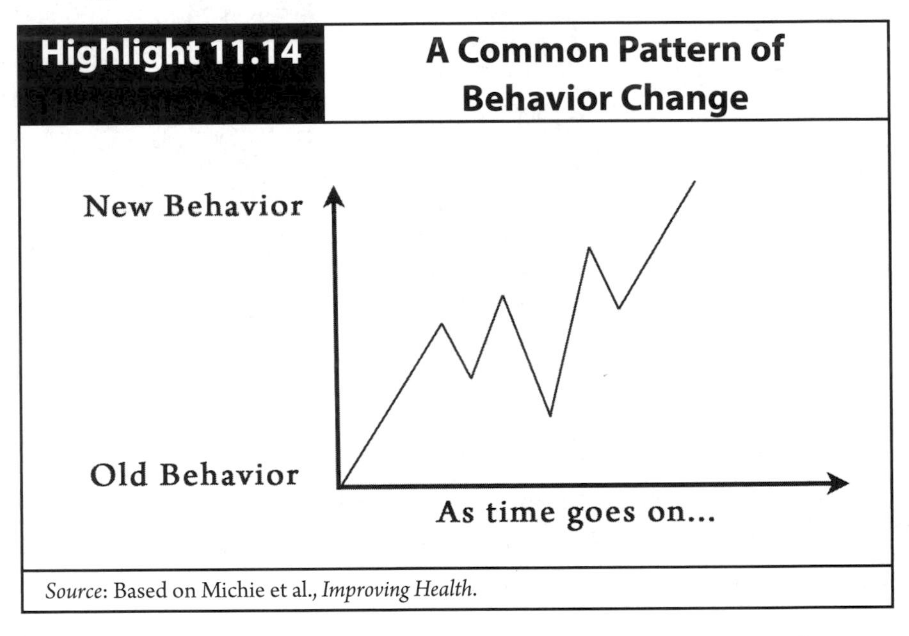

Source: Based on Michie et al., *Improving Health.*

If after making adjustments you are still not making progress, consider using the stepped approach mentioned before. That is, look for self-help materials or a support group to help you. If that doesn't work, consider using more professional help. These approaches are discussed later. Don't be too impatient. Remember it may take months before what you are trying to do becomes habitual or is otherwise achieved. Keep persevering. As the old saying goes, "If you don't succeed at first, try, try, again"—and learn from your experience!

Highlight 11.15 Perseverance and Success: Some Notable Quotes

"I've failed over and over and over again in my life and that is why I succeed."
(Michael Jordan)

"Success is not final, failure is not fatal: it is the courage to continue that counts."
(Winston Churchill)

"Success does not consist in never making mistakes but in never making the same one a second time."
(George Bernard Shaw)

"And will you succeed? Yes indeed, yes indeed! Ninety-eight and three-quarters percent guaranteed!"
(Dr. Seuss)

Sources: http://www.brainyquote.com/quotes/topics/topic_success.html and http://www.goodreads.com/quotes/tag/success, April 18, 2013.

Step 8: Maintain Your Gains

- Maintain your changed behavior and desired outcomes.
- Consider your actions a lifestyle change.
- Stop if your goal is a onetime event.

During this last step you should either try to maintain the gains that you have made or if you have achieved a one-time event, such as securing a new job or obtaining a divorce, you can stop your goal-related activity. Many types of behavior change and outcomes, however, are meant to be ongoing such as stopping smoking, gambling, and drinking alcohol; losing weight; managing diabetes; controlling anger; and maintaining good relations with a special person. If so, the habits you develop to achieve your goal need to become a lifestyle change.

For example, in my own case, I have reduced my weight to a desired level, but still generally continue to do what I did to achieve that goal. That is, I weigh myself regularly and, when I notice that my weight is going up slightly, I more carefully use the techniques that led to my earlier success. All of this is a lifestyle change that is satisfying and fairly easy to maintain because it is now habitual.

You too can have similar success reaching your own goals for changing your behavior. But remember, as indicated before in Highlight 11.2, "Change isn't an event; it's a process," and "People do not fail at self-modification because the techniques don't work; they fail because they don't use the techniques." Continue the process of using the techniques that work for you, as needed, after you have reached an ongoing goal that you value. Don't stop doing what works. Continue to make changes when needed. Keep up the process!

A Checklist You Can Use

The major ideas of the steps to behavior change presented are summarized in the checklist shown in Appendix 5. You may want to use that checklist to help ensure that you consider all of the techniques we have discussed to successfully achieve your goal.

Additional Resources Available to Help You Change

Self-Help Groups

Resources available to help you change include a number of clearinghouses and self-help groups. Highlight 11.16 indicates some of these. You may be surprised at how many local self-help groups there are. Tens of thousands exist around the world! Many are linked to broader organizations. For example, one sourcebook lists the following major self-help organizations and online resources:

- 30 dealing with *abuse* (child, sexual, and domestic spouse abuse; and violence),
- 77 dealing with *addictions* (including drug, alcohol, food, gambling, overspending, and sex addition),
- 92 dealing with *disabilities* (including amputation, autism, brain injury, hearing disorders, and mental retardation),
- 76 dealing with *family and parenting* (including adoption, foster families, marriage, general parenting, pregnancy and childbirth, separation and divorce, single parenting, stepparenting, and multiple births),
- more than 400 dealing with *health conditions* (including AIDS, Asthma, Undiagnosed Illness, Women's Health—and lots more!),

- 57 dealing with *mental health* (including, anxiety, depression, families of the mentally ill, mental health patients, schizophrenia, and self-injury), and
- others in *a miscellaneous category*, dealing with issues ranging from accident and crime victims, to caregivers of the elderly, prejudice, sexual orientation, speech and stuttering, trauma, and women's and men's issues.[40]

Such self-help resources are often associated with smaller local self-help groups, sometimes numbering into the thousands, depending on the particular area of interest. For example, as Highlight 11.16 indicates, Alcoholics Anonymous has over 100,000 local groups that meet worldwide, whereas Procrastinators Anonymous is a much smaller, internet-based support group.

Highlight 11.16 Examples of Self-Help Resources

Clearinghouses

American Self-Help Group Clearinghouse: a database of more than 1000 self-help organizations and clearinghouses worldwide (look for it online)

National Mental Health Consumers' Self-Help Clearinghouse: a mental health consumer assistance center (http://www.mhselfhelp.org)

Support Clearinghouse Directory: a U.S. and international directory to self-help groups (http://www.pkids.org/files/pdf/phr/12-03supportdirectory.pdf)

SAMHSA's National Registry of Evidence-based Programs and Practices: a registry of interventions concerning mental health promotion, substance abuse prevention, and treatment (http://www.nrepp.samhsa.gov/)

Organizations and Groups

Target	Organization/Group
Bereaved parents, family members, and their friends	The Compassionate Friends (600 chapters)
Survivors of child sexual abuse 18 years of age or older	Survivors of Incest Anonymous (300 groups)
Heart disease patients and their families	Mended Hearts (280+ chapters)

Current and former psychiatric patients and others struggling with emotional and mental health problems, including depression and anxiety	Recovery International (500+ groups)
Alcoholics	Alcoholics Anonymous World Services, Inc. (AA; 100,000 groups)
Families and friends of problem drinkers	Al-Anon Family Groups, Inc. (24,000+ groups)
People who wish to lose weight	Weight Watchers (96,000 regularly held meetings in 30 countries)
People who have a problematic relationship with food	Overeaters Anonymous (6,500 groups)
Victims of stroke, families, and friends	Stroke Clubs International (900+ clubs)
Addicts who wish to stop using drugs	Narcotics Anonymous (more than 60,000 meetings weekly in 129 countries)
Persons with a desire to stop compulsive gambling	Gamblers Anonymous® (2,400 chapters)
Persons wishing to stop addictive sexual behavior	Sex Addicts Anonymous® (700+ groups)
Individuals who wish to abstain from addictive or compulsive activities	SMART Recovery® (600+ affiliated groups)

Internet Support Groups and Resources

Target	Web Site/Support Group
Overweight individuals	ChooseMyPlate.gov
Adolescents & young women with eating disorders, poor body image, low self-esteem	Student Bodies©
Chronic procrastinators	Procrastinators Anonymous
People seeking to cope with a health problem or concern, such as AIDS and cancer; and caregivers of Alzheimer's patients	CHESS™ (Comprehensive Health Enhancement Support System)

Persons with cancer and their parents, other family, friends, and caregivers	ACOR (Association of Cancer Online Resources)
Various psychological conditions	psychcentral.com
Persons facing mental health and life-style choices concerning adults, families, children, and seniors	www.helpguide.org
Persons seeking information about psychological self-help	www.psychologicalselfhelp.org
Those wishing to improve their communication, public speaking, and leadership skills.	Toastmasters International (with 15,000 clubs in 142 ciountries) www.toastmasters.org
Major Sources: Yalom and Leszcz, *Group Psychotherapy*; White and Madara, *Self-Help Group Sourcebook*; American Self-Help Clearinghouse (online); Norcross et al., *Self-Help That Works*.	

Local self-help groups consist of people who share a problem, predicament, or goal and come together to help one another. Instead of self-help groups, they might better be called "mutual help" or "mutual support groups," since group members help one another. These groups are usually voluntary and nonprofit, and are run by their members. In addition to providing social support, they provide practical information and education based on the experiences of their members, help to develop coping skills, and may sometimes invite professionals to share their knowledge and insights.[41]

If you are interested in changing your behavior and need help, look online to locate relevant groups and materials. Use one of the clearinghouses or other resources mentioned in Highlight 11.16. Or do a general search of self-help groups and clearinghouses using keywords relevant to your interest. You will probably find a group close to you, especially if you live in a more developed country. You may also find a helpful online service with relevant articles and perhaps a chat room to discuss problems and ideas with others. Most of these services are free.

Highlight 11.17 Countries and Regions Having Self-Help Clearinghouses

Argentina, Australia, Austria, Belgium, Canada, Croatia, Denmark, England, Finland, Germany, Greece, Hungary, Ireland, Israel, Japan, Liechtenstein, Mexico, Netherlands, Norway, Poland, Romania, Scotland, Slovenia, Spain, Sweden, Switzerland, U.S.A, Wales.

Sources: American Self-Help Clearinghouse, online, October 2014; White and Madera, *Self-Help Group Sourcebook*.

Self-Help Books

Bookstores are filled with shelves of self-help books. However, be careful. Some books are based on scientific findings and have been shown to be effective. Others are only based on the assumptions, ideas, and opinions of their authors rather than on solid evidence of effectiveness. Some self-help books are useful, some have no effect, and a few are even harmful. For example, people have been made to feel as if they are failures when they did not achieve what a book indicated its readers should have been able to do. Even when self-help books have led to benefits, improvements often can be because of placebo effects, rather than any solid advice contained in the book.

Highlight 11.18 The Placebo Effect

Typically, 30 percent to 40 percent of people benefit from techniques mentioned in self-help books *if they expect or hope those techniques will work*. Such is the power of belief! Expectations of improvement can lead to improvement in many cases, even if a technique, medication, or procedure has no real causal effect! These placebo effects are real. They are akin to self-fulfilling prophecies. And the effects of many self-help books as well as many psychological treatments are based upon them.

Here are some related definitions from the *APA Dictionary of Psychology* by VandenBos:

- **placebo effect** a clinically significant response to a therapeutically inert substance or nonspecific treatment. It is now recognized that placebo effects accompany the administration of any drug (active or inert) and contribute to the therapeutic effectiveness of a specific treatment.
- **self-fulfilling prophecy** a belief or expectation that helps to bring about its own fulfillment, as...when a teacher's preconceptions about a student's ability to influence the child's achievement for better or worse.

And so, what to do? If you wish to buy a self-help book, here are some suggestions:

- *Don't buy a book that lacks references or a bibliography in the back.* It indicates that the author did not consider and cite research evidence available.
- *Don't buy a book simply because a celebrity endorses it or because it is on a best seller list.* Such endorsement and fame may be a result of good marketing aimed at making profits.
- *Don't buy a book that promises "fast and easy" results.* Such talk is probably a false promise. It may be marketing "hype."
- *Do buy self-help books recommended by reputable and knowledgeable organizations,* such as the organizations listed in Highlight 11.16. The book by John Norcross and others, *Self-Help That Works,* is also a useful guide.

In other words, be cautious when buying self-help materials.

Highlight 11.19 lists some self-help books shown to be effective, based on research evidence. These can be obtained without concern for their usefulness, since they are known to be helpful. If you are not sure what to buy, look at the clearinghouses, organizations and groups mentioned in Highlight 11.16, since they seem to provide a solid basis for obtaining helpful ideas, support, and guidance.[42] John Norcross's book, *Changeology,* lists recommended self-help books based on ratings by clinical and consulting psychologists. A more extensive listing of self-help books as well as autobiographies, films, online programs, internet sites, and self-help support groups for 41 behavioral disorders and life challenges such as aging, divorce, marriage, and pregnancy, is in *Self-Help That Works,* by John Norcross and others.

Highlight 11.19	**Some Self-Help Materials of Demonstrated Effectiveness**
Behavior Area	**Useful Materials**
Depression	• *Feeling Good: The New Mood Therapy*, by Burns • *Control Your Depression*, by Lewinsohn and others
Eating disorders	• *Overcoming Binge Eating*, by Fairburn
Insomnia	• *The Relaxation Response*, by Benson • *Relief from Insomnia*, by Morin • *:60 Second Sleep Ease: Quick Tips to Get a Good Night's Rest*, by Currie and Wilson
Panic disorders	• *Coping with Panic*, by Clum
Sexual dysfunctions	• *For Yourself: The Fulfillment of Female Sexuality*, by Barbach • *Becoming Orgasmic: A Sexual and Personal Growth Program for Women*, by Heiman and LoPiccolo • *Prolong Your Pleasure*, by Zeiss and Zeiss
Social phobia/social anxiety	• *Dying of Embarrassment: Help for Social Anxiety and Phobia*, by Markway and others • *A New Guide to Rational Living*, by Ellis and Harper
Specific phobias, such as to spiders, snakes, and others	• Various materials, including "CAVE," a computer-assisted vicarious exposure program
Source: Watkins and Clum, *Handbook of Self-Help Therapies*.	

Professional Help

If you follow a stepped approach to self-improvement, you will first try to change your behavior by using the steps and techniques described in this chapter. Then, if you decide that more help is needed, proceed to using self-help groups and other materials. If that doesn't work and you really want to change, consider obtaining professional help. Professionals will not solve your problems, but they can help you to solve them.[43] For example, if you can't develop enough commitment to change a problem behavior, a professional who uses the "Motivational Interviewing" approach may be able to help you. Such specialists may also be able to help you clarify your goals, develop a change plan, and support your change efforts.[44]

Mental health professionals licensed by many states include psychologists, psychiatrists, clinical social workers, counselors, psychiatric nurses, and marital/ family therapists. But be careful. In most places, nearly anyone can call him/ herself a "psychotherapist" or "therapist." Therefore, if you are looking for professional help, avoid using people who advertise themselves as such but are not officially licensed to practice by your state's government.[45]

Two good websites concerning the nature of therapy and how to select a psychotherapist are "http://aboutpsychotherapy.com" and "http://metanoia.org". Norcross's book on *Changeology* also has a useful Appendix titled "Selecting the Right Psychotherapist for You."

Do not feel that you have to be crazy to use a mental health professional! Ordinary, normal people having difficulty changing their behavior also use them too as guides and helpers. Perhaps the term "assisted self-help professional" is better to use than "mental health professional." In any case, seek professional help if needed, but approach the selection of such a helper carefully rather than casually.

How Others Try to Influence What You Do

It's More Than You May Realize

As an autopoietic being, no one else can really change your behavior except by overwhelming force that overcomes your control systems. Otherwise, the best they can do is try to influence what you do—and people are doing that all the time! From requests and commands to the subtle influences of advertising and propaganda, other people are constantly trying to affect your behavior. It seems that you face hundreds of attempts to influence you every day—perhaps 400 or more![46] For example:

- A family member, friend, co-worker, or even a stranger may ask you to do something such as, "Please pass the salt," "Let's go to a movie," or "Can you help me lift this?"
- Businesspeople try to persuade you to buy their products and services, usually so they can make a profit.
- The management of your employer and your immediate supervisor, if you have one, try to have you behave in ways that help them achieve their own goals.

- Governments try to gain support for their policies and actions, such as going to war.
- Politicians try to gain support for what they do and urge you to vote for them.
- The media is constantly trying to shape your perceptions. Think about it; most of what you know about the world is a result of media such as TV, radio, newspapers, magazines, and books. Such media are put together, in part, to make profits or to spread the word about causes that their producers are especially interested in. Examples include commercial television and talk radio.
- Groups and organizations with special interests, such as environmentalists, nonprofit agencies, schools, and religious organizations, try to influence how you think and act, and may try to solicit you for donations or other contributions.
- Con men and women, scam artists, and thieves may try to manipulate you into taking action that helps them, but not you, such as having you reveal the pin number to your bank account.

Highlight 11.20	**Attempts to Influence You and Others**

The world is awash with attempts to influence: blandishments of advertisers; twists of politicians; systematic campaigns of conversion, seduction, or trickery. Knowledge of how influence works will allow us to better resist it.

(Tharp, *Delta Theory*, 3-4)

Appendix 6 summarizes many of the ways that others try to influence you and what you do. As you look at it, think about your own experience. You are constantly exposed to efforts that attempt to influence your behavior. Many of these efforts have been successful—perhaps to your increasing dismay as you realize what has been happening. Sometimes, though, you may welcome the techniques that others use, especially if they help you to achieve your goals. However, if you realize that you are being manipulated in ways not to your liking, an awareness of techniques used by others, such as shown in Appendix 6, can help you to counteract those attempts. Techniques of influence range from the simple, such as a smile, to the ubiquitous, like advertising, to the legal, as laws, to architectural and infrastructure development, such as the construction of bike paths to encourage bike riding.

Major ways others try to influence what you do include efforts that attempt to:

- Have you perceive things as they wish you to perceive them.
- Direct your attention to areas they wish you to perceive.
- Create in your mind certain preferred states, goals, values, and other references that they want you to have.
- Help you perceive real or imagined error signals that you might otherwise not notice.
- Increase or decrease the importance of your error signals and related feelings about people, things, and events.
- Develop your knowledge, beliefs, skills, attitudes, and abilities in ways that lead to behaviors that they wish to either promote or discourage.
- Provide real or distorted feedback about areas that are of special interest to you, including what you do and how you are doing.
- Use social control methods, including legal, medical, and religious techniques, as well as personal approval and disapproval of you and what you do.
- Make it easier or harder for you to do some things.

They do these so that you will do what they want you to do—buy their products, vote in certain ways, or be a sucker or victim. They may also try to influence you to be a good person, worker, soldier, citizen, parent, or child. Even when they appear to do these things to help you, as you now know, they do so in order to achieve the goals and perceptions that they wish to achieve. That is why your parents often did things to affect your behavior, and you do too if you have children. That is why religious leaders want you to behave appropriately—to save your soul or otherwise lead a better life and benefit after you die. If you think about it, that's why you help other people too—to achieve or maintain perceptions and goals that are important to you!

Influencing others often helps all involved. For example, such influence is necessary if society is to exist. However, sometimes, people try to influence you in ways that may benefit them but harm you. For example, the thief who manages to have you tell him or her the pin number of your bank account is not helping you. He or she wants your money!

Dealing with External Influences

What can you do about external influences on your behavior? How can you avoid other people manipulating what you do? You can try to be more aware of such influences and the effect that others have on you. Some of those effects may be desirable, and you may want to continue them—such as influences by a respected colleague, friend, parent, teacher, or counselor. Others may be undesirable, and you may wish to avoid, reduce, or discontinue their effect on you and what you do. You may then try to do things such as avoid the sources of those influences, such as by turning off your TV, or shifting to another channel or source of information; by not cooperating or associating with certain people; by moving to a different location; or working in a different environment.

What you do is up to you. By following the steps in this chapter, you should be able to deal with important external influences if you really want to and it is possible to do so. If you have difficulty, try following the stepped approach mentioned before. Use self-help groups and other resources. Then, if you need to, make use of an assisted self-help professional.

To Conclude

This book has given you a solid basis for knowing why you do the things you do. The ideas we have discussed apply to other people, too. You now have a sound foundation for understanding why other people do the things they do, too. As an autopoietic being, it is ultimately up to you whether or not you wish to continue doing what you are doing or you wish to change. If you choose to change, the insights you have gained by reading this book should be of use to you. In any case, how you deal with the world around you depends on you and how you are internally organized and structured. It depends on how you perceive the world and attempt to control those perceptions. Once you understand this, you gain power over what you do.

Further Reading

A good book and easy read, based on solid research, that parallels the approach presented in this chapter is:

- John C. Norcross, *Changeology: 5 Steps to Realizing Your Goals and Resolutions* (New York: Simon and Schuster, 2012).

For a scholarly book aimed at college students but whose ideas have broad applicability, see:

- David L. Watson and Roland G. Tharp, *Self-Directed Behavior: Self-Modification for Personal Adjustment,* 10th ed. (Belmont, CA: Wadsworth, 2014).

For a more technical discussion of ideas about Perceptual Control Theory and behavior change, see:

- Richard H. Pfau, *Using Perceptual Control Theory as a Framework for Influencing and Changing Behavior* (paper prepared for presentation at the Control Systems Group International Conference, Boulder, Colorado, July 17-21, 2012). Available at http://www.academia.edu/874110/Pfau_-_CSG_Conference_Paper_-_11_June_2012_-_with_jpg_figure.

For a lighter read concerning habits and habit change, look at:

- Charles Duhigg, *The Power of Habit* (New York: Random House, 2012).

An easily read, research-based book concerning willpower and ways to increase it is:

- Roy F. Baumeister and John Tierney, *Willpower: Rediscovering the Greatest Human Strength* (London: Penguin Books, 2011).

A good review of the benefits and often-neglected ease of changing one's environment in order to eliminate or reduce tempting impulses is:

- Angela L. Duckworth, Tamar Szabó Gendler, and James J. Gross, "Situational Strategies for Self-Control," *Perspectives on Psychological Science* 11, no. 1 (2016): 35-55.

Useful introductions to self-help books, support groups, internet resources, autobiographies, and films are provided by:

- The clearinghouses, organizations, and other resources mentioned in Highlight 11.16.
- John C. Norcross and others, *Self-Help That Works: Resources to Improve Emotional Health and Strengthen Relationships* (Oxford: Oxford University Press, 2013).

For a collection of inspiring stories and practical insights from people who have successfully made health behavior changes, see:

- Joseph Leutzinger and John Harris, *Why and How People Change Health Behaviors: A Collection of Success Stories,* vol. 1 (Omaha, NE: Health Improvement Solutions, 2005).

Good introductions to how people and organizations attempt to influence you and others, include:

- Robert B. Cialdini, *Influence: Science and Practice,* 5th ed. (Boston: Pearson, 2009).
- Christopher Hadnagy, *Social Engineering: The Art of Human Hacking* (Indianapolis: Wiley, 2011).
- Wayne Le Cheminant and John M. Parrish, *Manipulating Democracy: Democratic Theory, Political Psychology, and Mass Media* (New York: Routledge, 2011).
- Magedah E. Shabo, *Techniques of Propaganda and Persuasion* (Clayton, DE: Prestwick House, 2008).
- B. J. Fogg, *Persuasive Technology: Using Computers to Change What We Think and Do* (Amsterdam: Morgan Kaufmann, 2003).
- Max Sutherland, *Advertising and the Mind of the Consumer: What Works, What Doesn't, and Why,* rev. 3rd ed. (Crows Nest NSW, Australia: Allen and Unwin, 2008).
- Martin Lindstrom, *Brandwashed: Tricks Companies Use to Manipulate Our Minds and Persuade Us to Buy* (New York: Crown Business, 2011).

Endnotes

1. More technically we can say that behavior can be changed by changing any of the parameters of related control systems, including environmental inputs to those systems.

2. Baumeister, Heatherton, and Tice, *Losing Control*; Watson and Tharp, *Self-Directed Behavior*, 10th ed.

3. Leutzinger and Harris, *Why and How People Change*, 18.

4. This approach is akin to the Method of Levels, a psychotherapeutic approach based on Perceptual Control Theory during which a person is guided to consciously move his or her thinking up the hierarchy of control systems to higher levels as indicated by Highlights 3.7 and 7.4. Since, according to PCT, those higher levels set reference standards for lower levels including those that control what one does, identifying an important value or principle can help to resolve conflicts and take action consistent with that significant reference.

5. Miller and Rollnick, *Motivational Interviewing*, 11.

6. Watson and Tharp, *Self-Directed Behavior*, 9th ed., 37.

7. Action Theory argues that such information gathering is a prerequisite of planning. The action theory approach is consistent with that suggested by this chapter, including its emphasis on goal development, the collection of information, plan development and execution, monitoring of the execution of the plan, and feedback. For more information, see Frese and Zapf, "Action as the Core of Work Psychology"; Frese and Fay, "Personal Initiative"; and Raabe, Frese, and Beehr, "Action Regulation Theory."

8. Leutzinger and Harris, *Why and How People Change*, 8.

9. Watson and Tharp, *Self-Directed Behavior*, 9th ed., 70, 71. If you are academically inclined, look at "Chapter 3: Self-Knowledge: Observation and Recording," 69-108 and other parts of their useful book.

10. Ibid., 37, based on behavior engaged in by many overweight people.

11. Baumeister and Tierney, *Willpower*, 62, 173.

12. Locke and Latham, *Goal Setting*; Locke and Latham, *A Theory of Goal Setting*; Michie et al., "Effective Techniques in Healthy Eating," 690; Dombrowski and others, "Identifying Active Ingredients," 7, 21.

13. Michie et al., *Improving Health*, 31.

14. Baumeister and Tierney, *Willpower*, 38-39.

15. Norcross, *Changeology*, 87.

16. Watson and Tharp, *Self-Directed Behavior*, 9th ed., 104.; Michie et al., "Effective Techniques in Healthy Eating," 690; Dombrowski et al., "Identifying Active Ingredients," 7, 21.

17. Quinn, Pascoe, Wood, and Neal, "Can't Control Yourself," 499, 509-510.

18. John C. Norcross, *Changeology*, 58.

19. Moos, *The Human Context*, 398.

20. Gamow's definition of stress (i.e., "Stress is created by the gap between how the world is and how we want it to be") is compatible with the PCT concept of an error signal (i.e., that an error signal is the gap between our perception of the world and a related reference value for that perception). Error signals result in automatic reactions of our bodies, including secretion of the hormones cortisol and epinephrine that result in what is called the stress response. See Gamow, *Freedom from Stress*; Nelson, *Behavioral Endocrinology*, 4th ed., 581-587.

21. Sapolsky, *Why Zebras Don't Get Ulcers*, 384-418.

22. Fog, Black *Path Behavior Guide, Version 0.40*, 8, accessed March 12, 2012, from https://the-stanford-persuasive-tech-lab.myshopify.com

23. Gollwitzer, Bayer, and McCulloch, "The Control of the Unwanted," 485; Gollwitzer and Gabriele Oettingen, "Planning Promotes Goal Striving," 162-185.

24. For details see Gollwitzer, "Implementation Intentions."

25. Watson and Tharp, *Self-Directed Behavior*, 9th ed., 40.

26. Ibid.

27. Hadnagy, *Social Engineering*, 21.

28. This activity tracker was a "CR Best Buy" according to *Consumer Reports*, February 2016, 21.

29. Grilly and Salamone, *Drugs, Brain, and Behavior*.

30. Baumeister and Tierney, *Willpower*, 124-141.

31. Muraven, Baumeister, and Tice, "Longitudinal Improvement of Self-Regulation"; Gailliot, Plant, Butz, and Baumeister, "Increasing Self-Regulatory Strength"; Muraven, "Building Self-Control Strength."

32 Gailliot et al., "Self-Control Relies on Glucose;" Gailliot, Peruche, Plant, and Baumeister, "Stereotypes and Prejudice."

33. Cialdini, *Influence*, 5th ed., 71.

34. Norcross, *Changeology*, 99.

35. Watson and Tharp, *Self-Directed Behavior*, 10th ed., 258.

36. Ibid., 252.

37. Michie, Rumsey, and others, *Improving Health*, 37.

38. B. J. Fogg seems to differ from this idea, since he suggests dealing with three

different goals at once. But his purpose seems to have you learn from failure, because you will probably fail to achieve at least one of the goals you have set by diversifying your efforts. By doing so he hopes that you will begin to understand why things did not work out so that you will be able to design more effective habits for yourself the next time: Fogg according to www.oprah.com/spirit/The-Fastest-Way-to-Make-Change-How-to-Change-Your-Life/print/1, accessed March 8, 2013.

39. Watson and Tharp, *Self-Directed Behavior,* 10th ed., 45.

40. White and Madara, *The Self-Help Group Sourcebook*. Listings of these and other groups are now easily accessible online.

41. Silverman, "An Introduction to Self-Help Groups."

42. For a critique of the self-help movement, see Pearsall, *The Last Self-Help Book*.

43. Watson and Tharp, *Self-Directed Behavior,* 9th ed., 306.

44. Miller and Rollnick, *Motivational Interviewing.*

45. Norcross, *Changeology,* 220-222.

46. Brown, *Influence,* 3; Rhoads, "Everyday Influence," 1.

APPENDICES

Appendix 1: Levels of Analysis and Explanation

When explaining how you function, we can focus on any of several levels of analysis. These range from considering minuscule things such as genes and cells, to broad concepts such as countries and their policies. The levels considered in this book are:

- **The Broader Environment:** This refers to more distant surroundings and events that affect your immediate environment such as government policies, laws, general economic conditions, and technology. Theories and models at this level include Marxist-Leninist theory and Keynesian economics. This level is sometimes called the "macro-environment."
- **The Immediate Environment:** This level consists of surroundings and events that directly impinge on your senses such as your family, friends, news media heard or looked at, resources available, and settings like a school classroom, workplace, stores, and roadways. Theories and models at this level include behaviorism and family system theory. This level is sometimes called the "micro-environment."
- **The Person/Individual Level:** This is the level of the whole person and descriptive terms such as your sensations, feelings, emotions, attitudes, beliefs, desires, thoughts, intentions, purposes, plans, and activities. Psychological theories at this level include social cognitive theory and goal-setting theory, where properties of interest are of you, a person, not subsystems within you.[1]
- **The Sub-Personal Level:** This refers to internal systems such as your nervous system, respiratory system, and circulatory system. This level includes descriptive models and theories of interior processes and functions such as Perceptual Control Theory that focuses on the nervous system.
- **The Cell and Molecular Level:** This is the level of cells, genes, neurotransmitters, molecules, chemical actions, and related physical

and chemical statements. Physiological models and theories include neurobiological explanations of muscle activity and biochemical explanations of protein synthesis.

The levels above are mentioned for several reasons. The first is to point out that the analysis of your behavior in the first chapters of this book focuses on you and your body at the sub-personal and person/individual levels. Later chapters focus more on your immediate and broader environments. Secondly, the different levels are mentioned in recognition of the fact that a comprehensive view of your behavior should keep in mind that functioning at any one level is influenced by factors associated with other levels around it.[2] An understanding of these other factors can help you understand better why you do the things you do.[3] Finally, the different levels are presented to help point out that at each level different types of description and explanation are required.

Regarding the last reason mentioned, although the different levels of explanation can help us to understand human behavior better, problems sometimes arise when people attempt to apply concepts and terms that are applicable at one level to phenomena at another level, such as when they confuse why we do things at the person level with why our body moves the way it does at the physiological level of cells and molecules.[4] For example, the concepts of "purpose," "function," and "goal" are descriptive terms that lie in the domain of observers at the personal level and are considered unnecessary and misleading when applied to the mechanics of autopoiesis at the cell and molecular level according to Matura and Varela who coined the word autopoiesis.[5] Similarly, a description of the chemical mechanisms of neural cells and the interactions of those cells will not be of much use to understand purposes you have that affect your behavior or, for example, why you shop at one store rather than another.

Highlight A.1 Differing Levels of Analysis Can be Complementary

Referring to teleological descriptions at the person/individual level, and mechanical descriptions at the physiological, personal, and cell-molecular levels, Harry Binswanger writes, "The two approaches are complementary: one gives a general overview which needs to be supplemented by specific detail, the other gives specific detail which needs to be supplemented by a general overview."

(Binswanger, *Biological Basis,* 196)

Appendix 2: Guide for Analyzing a Current Behavior

To analyze and better understand something that you are doing, copy this guide, consider each of the following questions and write your answer in the space provided.

1. What are you doing?

2. What are you trying to control?

3(a). What was your perception of what you are dealing with before you started doing what you are doing?

3(b). What is your perception of what you are dealing with now?

4. What is your reference for what you are doing?

5(a). What error was occurring or anticipated <u>before</u> you began doing what you are doing?

5(b). What is the error like <u>now</u>?

6. What else in the environment is affecting what you are trying to control?

7. What side effects is your behavior having?

If you wish to analyze more, additional probing questions are:

8. What influences are there on your input function and resulting perceptions?

9. What influences are there on the references you are using?

10. What internal influences are there on what you are doing?

11. What other immediate environmental conditions are affecting you and what you are doing?

12. What broader environmental influences are there on what you are doing and its effects?

Appendix 3: Guide for Analyzing a Past Behavior

To analyze and better understand something that you did in the past, copy this guide, consider each of the following questions and write your answer in the space provided.

1. What did you do?

2. What were you trying to control?

3(a). What was your perception of what you were dealing with before you did what you did?

3(b). What was your perception of what you were dealing with after you did what you did?

4. What was your reference for what you did?

5(a). What error was occurring or anticipated before you did what you did?

5(b). What was the error like after you did what you did?

6. What else in the environment affected what you were trying to control?

7. What side effects did your behavior have?

If you wish to analyze more, additional probing questions are:

8. What influences were there on your input function and resulting perceptions?

9. What influences were there on the references you used?

10. What internal influences were there on what you did?

11. What other immediate environmental conditions affected you and what you did?

12. What broader environmental influences were there on your behavior and its effects?

Appendix 4: Planning Guide for Behavior Change

This guide is provided for you to copy and write your plan in the spaces provided.

My Goal (Be sure that your goal is specific, measurable, achievable, relevant, and time bound):

Major Behaviors Involved to reach your goal:

How I Will Monitor Progress
[a] Toward my goal:

[b] Toward carrying out this plan:

Environmental Changes I Will Make:

To increase desirable behavior, such as to use cues, signs, reminders; remove obstacles; make the behavior easier to do; provide yourself with useful resources; use other people; change your location:

To reduce undesirable behavior, such as to eliminate or avoid undesirable triggers; remove objects needed to perform the behavior; make it harder to do the behavior:

How I Will Deal with Undesirable Triggers, Temptations, and Difficult Situations. List your if-then plans:

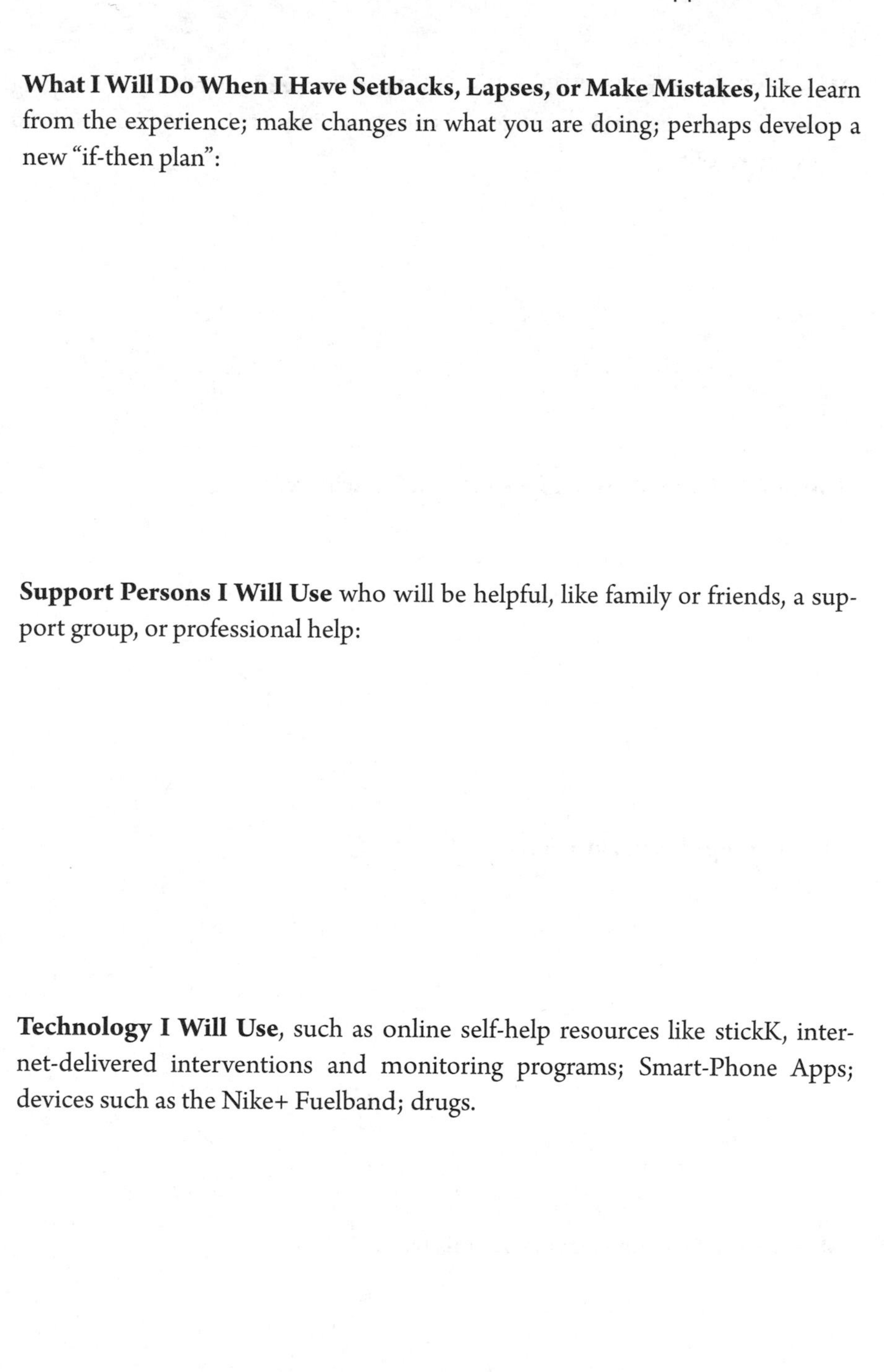

What I Will Do When I Have Setbacks, Lapses, or Make Mistakes, like learn from the experience; make changes in what you are doing; perhaps develop a new "if-then plan":

Support Persons I Will Use who will be helpful, like family or friends, a support group, or professional help:

Technology I Will Use, such as online self-help resources like stickK, internet-delivered interventions and monitoring programs; Smart-Phone Apps; devices such as the Nike+ Fuelband; drugs.

How I Will Ensure That I Have the Ability to Succeed, such as by developing needed knowledge, skills, confidence, and commitment; by obtaining needed tools and resources:

How I Will Reward Myself for small and big achievements:

Other Things I Will Do, if any:

My Starting Date for carrying out this plan: ______________________

Appendix 5: Behavior Change Checklist

Copy and use this checklist to record your progress and remind yourself what to do.

Step 1: Decide to change and mentally commit yourself to doing so

- ☐ Consider why you are dissatisfied.
- ☐ Clarify to yourself what you want to do differently.
- ☐ Commit yourself to doing what is required.

Step 2: Decide on a general goal to achieve.

- ☐ My general goal is to ____

Step 3: Gather information.

- ☐ Obtain information about yourself, situations, and people related to your goal.
- ☐ Obtain information about how others have dealt with similar behavior change.

Step 4: Clarify your goal.

- ☐ Your goal should be specific, measureable, attainable, relevant, and time bound.

Step 5: Create a plan to achieve your goal.

- ☐ Specify the behaviors involved.
- ☐ Decide how you will monitor progress.
- ☐ Decide what environmental changes you need to make.
- ☐ Consider how you will avoid temptations and other undesirable triggers.
- ☐ Develop "if-then" plans to deal with threatening and opportune situations.
- ☐ Prepare to deal with slips, errors, and lapses.

- ☐ Consider using support persons and groups.
- ☐ Consider using technology.
- ☐ Ensure that you have the knowledge, skills, commitment, and resources needed.
- ☐ Consider using a public contract and penalties, such as stickK.
- ☐ Plan to reward yourself along the way.
- ☐ Write down your plan by using Appendix 4 as a guide.
- ☐ Specify your starting date.

Step 6: Carry out your plan.

- ☐ Perform the behaviors involved:
  - ○ Avoid willpower depletion by having enough sleep, good nutrition, and not too many goals and targets at one time.
  - ○ Build habits.
- ☐ Be sure to monitor progress toward your goal and implementation of your plan.
- ☐ Make environmental changes that are needed.
- ☐ Avoid temptations and other undesirable triggers.
- ☐ Carry out your "if-then" plans to deal with threatening and opportune situations.
- ☐ Deal with slips, errors, and lapses.
- ☐ Use your support persons and groups.
- ☐ Access and use technology, if appropriate.
- ☐ Develop the knowledge, skills, resources, and commitment needed.
- ☐ Reward yourself along the way.

Step 7: Persevere and Revise Your Plan as Needed.

- ☐ Expect and deal with lapses, mistakes, and failures.
- ☐ Be flexible; revise your plan when needed.

- ☐ If what you are doing is not succeeding, use a stepped approach.

Step 8: Maintain Your Gains.

- ☐ Maintain your changed behavior and desired outcomes.
- ☐ Consider your actions a lifestyle change, or
- ☐ Stop if your goal is a onetime event.

Appendix 6: Some Ways Others Attempt to Influence Your Behavior

A. Your Perceptions

Others often try to affect how you perceive the world by using techniques such as the following. They do this knowing that what you perceive can influence what you do.

Directing your attention, such as by means of the intensity, movement, repetition, contrast, novelty, troubling nature, or sexual content of things perceived, for example, as TV does.

Cues, prompts, and other nudges aimed at getting you to act in certain ways.

Framing and spin techniques to affect the meaning of your perceptions.

Building rapport with you by expressing genuine interest in you, dressing appropriately, letting you do most of the talking, listening well, reflecting empathy with you, matching your body language, and matching your vocal tone, speech pattern, and breathing rate.[a]

Advertising/marketing, used to shape your perceptions of products, services, organizations, and people.

Social marketing and cause marketing that seeks to change personal behavior or public policy by using advertising and marketing techniques to effect social change.

Formal education, at schools, colleges, and universities.

Training and development, at workplaces, including induction, orientation, staff development programs, and apprenticeships.

Perception management, as used by governments, other organizations, and individuals including politicians.

Propaganda, which often overlaps with perception management.

Drugs and other medical techniques, including alcohol and anti-depressants.

(a) Christopher Hadnagy, *Social Engineering: The Art of Human Hacking,* 162-171.

B. Your References

Others may try to stimulate or establish goals, desires, plans, and other references that they want you to have and act upon, by using techniques such as the following.

Assigning goals, such as when a parent tells you to do something or when a teacher or manager sets a goal for you to achieve.

Exposing you to what other people are doing, saying, or thinking.

Providing information about expected behavior, norms, standards, or procedures—such as by standard operating procedure manuals, job descriptions, religious commandments and sins, and by persuasion techniques.

Creating a sense of duty, such as by military training, socialization at home, religious indoctrination, and other social control techniques.

Changing the importance of reference signals, such as via hormones/drugs, psychotherapy, meditation techniques, motivational interviewing, and team competition.

Priming your goals and other reference values by advertisements, announcements, news articles, signs, cues, music, and placement of products.

Helping you identify your higher order and lower order goals, such as during executive coaching and some types of psychotherapy.

Forms of psychotherapy by which the patient and therapist agree on the goals of treatment as well as tasks and/or plans to achieve those goals.

Establishing or changing laws, regulations, policies, and taxes that establish norms and consequences for certain behaviors.

Helping with or otherwise having you develop plans for achieving desired goals, such as by means of action planning, implementation intentions, "if-then" plans, and setting graded tasks.

C. Your Error Signals

Others may try to influence error signals that you experience, knowing that such signals may stimulate you to action.

Making you aware of error signals that may occur or that you should be experiencing, for example, by asking in a TV commercial, "Do you have stinky breath?"

Increasing or decreasing the importance of error signals you may be or should be experiencing, for example, "Did you know that smoking may kill you!"

D. Your Actions

Others may also do things the following to stimulate, develop, impair, or otherwise influence your behavior and ability to behave in ways that they desire.

Developing knowledge and skills you need in order to achieve goals desired by the influencer, for example, by:

- Teaching/instructing you or arranging relevant experience, via schooling, training, coaching, behavioral rehearsal, or roleplaying in a simulated setting—for instance, at work, to develop stress management, time management, or other skills.
- Providing self-study options and programs.
- Having you observe other people; showing what other people are saying and doing via the media and advertisements.

Affecting your selection of desired behavior by:

- Priming you with advertisements, speeches, news articles, and propaganda.
- Providing environmental cues, signs, reminders, prompts, nudges, and default selections.
- Providing support and encouragement from others.

- Providing relevant information about the consequences of behaviors and the causes of certain behavior, opportunities and resources available; by using advertising and marketing techniques; and by consultation or therapy.
- Helping you do a cost-benefit analysis of changing and not changing your behavior.
- Evoking positive and negative feelings about alternative behaviors.
- Increasing your self-efficacy/confidence in changing and successfully carrying out certain behavior.
- Increasing your self-control/willpower, by having you exercise self-control and willpower over time.
- Using the Premack principle, where you have to do something first in order to perform a behavior you want to do.
- Providing or developing a structure to guide decisions and behavior, such as planning tools and programs, decision trees, and suitable architecture.

Changing your output quantities by affecting your stamina and/or the intensity, extent, or continued production of certain behavior, by:

- Arranging or providing opportunities for prior exercise and practice, including physical practice and your practice of self-control.
- Bolstering or decreasing your self-control resources.
- Making nutritional changes; providing glucose supplements.
- Drugs such as for ADHD, depression, and other mental disorders; caffeine effects on muscular work and sustained intellectual effort; disinhibition effects of alcohol.
- Affecting your expectancies.

E. Feedback to You

Others may arrange or provide you with information about how well you are doing, knowing that such feedback can affect what you do.

Using informal social control, such as praise, compliments, corrections, and frowning.

Grades, awards, public recognition, including stars and smiley faces for schoolchildren, "employee of the month" awards, and employee bonuses.

Coaching, that may also include suggestions for improvement.

Performance appraisal, including 360-degree feedback at workplaces, and using mystery shoppers or diners at businesses.

Arranging regular practice, such as monthly police officer practice at a pistol range.

Briefing meetings to keep you informed of progress.

Comparing your progress to other people as a parent or supervisor may do.

Using technology such as hybrid car dashboard displays that indicate fuel efficiency, vehicle activated speed signs, biofeedback, GPS devices, buzzers, computer warnings, pedometers.

Program evaluations to obtain information about how goals are being achieved and what needs to be changed.

F. Your Environment

Others may try to affect you by using your environment to trigger your habits and other behavior, affect your learning, or otherwise influence what you do in ways such as those shown below.

All of "A" to "E" above, especially the following:

The use of social control methods such as those listed in Highlight 4.10 and briefly summarized below.

- **Informal techniques:** pats on the back, smiles, kisses, ostracism.
- **Organizational techniques:** provision of orientation and training, feedback, use of local opinion leaders.
- **Medical techniques:** drug and alcohol rehabilitation programs, omega-3 fatty acid diet supplements.
- **Legal techniques:** laws; rewards, such as tax breaks and deductions; imprisonment, exile; and death.
- **Religious and supernatural techniques:** commandments, moral precepts, rewards and punishments that individuals receive upon their death.[a]

Increase chances of the desired behavior occurring by:

- Priming your perceptions, reference values and/or behavior.
- Promises and signed behavioral contracts; by creating obligations .
- Incurring reciprocation, as Highlight 6.8 explains.
- Using default choices since people tend towards the status quo.
- Stress management/emotional control training.
- Use of technology, including forcing functions such as forcing you to fully close a microwave door before it will operate, computer and object simulations, virtual reality therapy, mobile phones, speed monitoring awareness trailers, surveillance cameras, and computer monitoring techniques at work.[a]
- Making it more difficult or costly to carry out other behavior.

Facilitating your behavior by:

- Removing temptations such as unhealthy food from easy access.
- Providing checklists, signs, cues, prompts, reminders, color coding, information, helpful tools and technologies like calculators, computer programs, GPS systems.[(b)]
- Guiding or encouraging desired behaviors using suitable architecture, facilities and infrastructure like bike paths to encourage bike use.
- Providing convenient access to resources, such as nearby health clinics, condoms, free needles, and insurance to encourage safer health practices.
- Providing social support such as counseling and other social services.
- Providing welcome kits to new residents that contain free coupons, passes, tickets, and discounts.

Removing obstacles interfering with or preventing desired behaviors or performance, for example:

- At the workplace, obstacles may include lack of time to do what you want to do, lack of authority, lack of tools, lack of information including information about job expectations and feedback about what you are doing, a shortage of resources, too many distractions, poor lighting, uncomfortable surroundings, intrusion of too many phone calls, or less important but more immediate problems.[(b)]
- Providing childcare, if you are a parent, so you can work.
- Providing public transport to provide access to work and shopping for people without cars.

(a) B. J. Fogg, *Persuasive Technology: Using Computers to Change What We Think and Do,* (Amsterdam: Morgan Kaufmann, 2003) is a good early introduction to some of these techniques. Fogg's book, *Mobile Persuasion: 20 Perspectives on the Future of Behavior Change* (Stanford, CA: Stanford Captology Media, 2007) focuses on mobile phones.

(b) Source: Robert F. Mager and Peter Pipe, *Analyzing Performance Problems: or You Really Oughta Wanna,* 2nd ed. (Belmont, CA: Lake Publishing, 1984); and the 3rd ed. (Atlanta: Center for Effective Performance, 1997).

Note: The techniques shown in this section overlap with those shown in sections "A" to "E" in part because any influence exerted by other people must be exerted by means of the environment around you. Even though these techniques overlap with those previously shown, they are presented here as an alternate way of viewing these influences.

Endnotes

1. Vancouver, "The Depth of History," 41.

2. Ford, *Humans as Self-Constructing Living Systems*, 40.

3. Vancouver, in his article "The Depth of History" provides good illustrations of this when he indicates how Perceptual Control Theory (at the sub-personal level) provides useful elaborations of Goal-Setting Theory and Social Cognitive Theory (at the person/individual level). Additional examples are provided in Chapter 9 of this book as part of the discussion of Perceptual Control Theory as a metatheory.

4. Rose, *Lifelines*, 95, 304-305; Dretske, *Explaining Behavior*, 36; Maturana and Varela, *Autopoiesis and Cognition*, 50. An extensive scholarly discussion of various conflations that lead to confusion and irresolvable, fruitless disputes is the book by Sandis, *The Things We Do*.

5. Maturana and Varela, *Autopoiesis and Cognition*, xix, 50, 86.

REFERENCES

Abraham, Charles, and Susan Michie. "A Taxonomy of Behavior Change Techniques Used in Interventions." *Health Psychology* 27, no. 3 (2008): 379-387.

Adolph, Karen E., Amy S. Joh, John M. Franchak, Shaziela Ishak, and Simone V. Gill. "Flexibility in the Development of Action." In *Oxford Handbook of Human Action*, edited by Ezequiel Morsella, John A. Bargh, and Peter M. Gollwitzer, 399-426. Oxford: Oxford University Press, 2009.

Agency for Toxic Substances and Disease Registry. "Lead: Health Effects." Atlanta: www.atsdr.cdc.gov/ToxProfiles/tp13-c3.pdf. Downloaded 3 December 2012.

Ainslie, George. *Breakdown of Will*. Cambridge: Cambridge University Press, 2001.

Alexander, Rudolf, Jr. *Human Behavior in the Social Environment: A Macro, National, and International Perspective*. Los Angeles: Sage, 2010.

American Self-Help Clearinghouse. *Support Clearinghouse Directory*. n.p.: Mental Health Network, 2005.

Ariely, Dan. *Predictably Irrational: The Hidden Forces That Shape Our Decisions*. rev. ed. New York: HarperCollins, 2009.

Aronson, Elliot. *The Social Animal*. 10th ed. New York: Worth Publishers, 2008.

Ashby, W. Ross. *Design for a Brain: the Origin of Adaptive Behavior*. London: Chapman and Hall, 1952.

Ashby, W. Ross. *Design for a Brain: The Origin of Adaptive Behavior*, 2nd ed. rev. London: Chapman and Hall, 1960.

Atkins, Robert C. *Atkins for Life: The Complete Controlled Carb Program for Permanent Weight Loss and Good Health*. New York: St. Martin's Press, 2003.

Baldwin, John D. "Habit, Emotion, and Self-Conscious Action." *Sociological Perspectives* 31, no. 1 (January 1988): 35-58.

Bandura, Albert. *Social Foundations of Thought and Action: A Social Cognitive Theory*. Upper Saddle River, NJ: Prentice Hall, 1986.

Bandura, Albert. *Self-Efficacy: The Exercise of Control*. New York: W. H. Freeman and Company, 1997.

Barbach, Lonnie Garfield. *For Yourself: The Fulfillment of Female Sexuality*. New York: Doubleday, 1974.

Bargh, John A. "Auto-Motives: Preconscious Determinants of Social Interaction." In *Handbook of Motivation and Cognition: Foundations of Social Behavior*. vol. 2. edited by E. Tory Higgins and Richard M. Sorrentino, 93-130. New York: Guilford Press, 1990.

Bargh, John A. "The Automaticity of Everyday Life." In *The Automaticity of Everyday Life: Advances in Social Cognition*, vol. 10, edited by Robert S. Wyer, Jr. 1-61. Mahwah, NJ: Lawrence Erlbaum Associates, 1997.

Bargh, John A. "Reply to the Commentaries." In *The Automaticity of Everyday Life: Advances in Social Cognition*, vol. 10, edited by Robert S. Wyer, Jr. 231-246. Mahwah, NJ: Lawrence Erlbaum Associates, 1997.

Bargh, John A. "What Have We Been Priming All These Years? On the Development, Mechanisms, and Ecology of Nonconscious Social Behavior." *European Journal of Social Psychology* 36 (2006): 147-168.

Bargh, John A. "Free Will Is Un-natural." In *Are We Free? Psychology and Free Will*, edited by John Baer, James C. Kaufman, and Roy F. Baumeister, 128-54. Oxford: Oxford University Press, 2008.

Bargh, John A., and Peter Gollwitzer. "Environmental Control of Goal-Directed Action: Automatic and Strategic Contingencies Between Situation and Behavior." In *Nebraska Symposium on Motivation: Integrative Views of Motivation, Cognition, and Emotion*. vol. 41. edited by William D. Spaulding, 71-124. Lincoln: University of Nebraska Press, 1994.

Barker, Roger G., *Ecological Psychology: Concepts and Methods for Studying the Environment of Human Behavior*. Stanford: Stanford University Press, 1968.

Barker, Roger G., and Herbert F. Wright. *Midwest and Its Children: The Psychological Ecology of an American Town*. Hamden, CT: Archon Books, 1971. First published 1955.

Barker, Roger G., and Associates. *Habitats, Environments, and Human Behavior: Studies in Ecological Psychology and Eco-Behavioral Science from the Midwest Psychological Field Station, 1947-1972*. San Francisco: Jossey-Bass Publishers, 1978.

Bateson, Gregory. *Mind and Nature: A Necessary Unity*. Cresskill, NJ: Hampton Press, 2002.

Baumeister, Roy F., Todd F. Heatherton, and Dianne M. Tice. *Losing Control: How and Why People Fail at Self-Regulation*. San Diego, CA: Academic Press, 1994.

Baumeister, Roy F., and Kristin L. Sommer. "Consciousness, Free Choice, and Automaticity." In *The Automaticity of Everyday Life: Advances in Social Cognition*, vol 10, edited by Robert S. Wyer, Jr., 75-81. Mahwah, NJ: Lawrence Erlbaum Associates, 1997.

Baumeister, Roy F., and John Tierney. *Willpower: Rediscovering the Greatest Human Strength*. New York: Penguin Books, 2011.

Becker, Jill B., S. Marc Breedlove, and David Crews, eds. *Behavioral Endocrinology*. Cambridge, MA: MIT Press, 1992.

Becker, Jill B., S. Marc Breedlove, David Crews, and Margaret M. McCarthy, eds. *Behavioral Endocrinology*. 2nd ed. Cambridge, MA: MIT Press, 2002.

Bennhold, Katrin. "In Sweden, the Men Can Have It All." *New York Times* (June 9, 2010): downloaded June 11, 2010.

Benson, Herbert. *The Relaxation Response*. New York: Harper, 1975; rev. ed., HarperCollins, 2000.

Berger, Jonah, Marc Meredith, and S. Christian Wheeler. "Contextual Priming: Where People Vote Affects How They Vote." *PNAS* 105, no. 26 (1 July 2008): 8846-9.

Berger, Kathleen Stassen. *The Developing Person: Through the Life Span*. 6th ed. New York: Worth, 2005.

Bernstein, Douglas A., E. J. Roy, T. K. Srull, and C. D. Wickens. *Psychology*. 2nd ed. Boston: Houghton Mifflin, 1991.

Biddle, Bruce. *Role Theory: Expectations, Identities, and Behaviors*. New York: Academic Press, 1979.

Binswanger, Harry. *The Biological Basis of Teleological Concepts*. Los Angeles: Ann Rand Institute Press, 1990.

Bogenschneider, Karen. *Family Policy Matters: How Policymaking Affects Families and What Professionals Can Do*. 2nd ed. Boca Raton, FL: CRC Press, 2006.

Bohannon, John. "The Theory? Diet Causes Violence. The Lab? Prison." *Science* 325 (25 September 2009): 1614-6.

Bohner, Gerd, and Michaela Wänke. *Attitudes and Attitude Change*. East Sussex, UK: Psychology Press, 2002.

Bossel, Hartmut. *Systems and Models: Complexity, Dynamics, Evolution, Sustainability*. Norderstedt, Germany: Books on Demand GmbH, 2007.

Boston University Medical Center. "Alcohol Taxes Protective Against Binge Drinking, Study Shows." *Science Daily* (January 5, 2015): http://www.sciencedaily.com/releases/2015/01/150105141705.htm.

Bourbon, W. Thomas. "Invitation to the Dance: Explaining the Variance When Control Systems Interact." *American Behavioral Scientist* 34, no. 1 (September/October 1990): 95-105.

Bourbon, W. Thomas. "On the Accuracy and Reliability of Predictions by Perceptual Control Theory: Five Years Later." *The Psychological Record* 46, no. 1 (Winter 1996): 39-47.

Brizendine, Louann. *The Female Brain*. New York: Three Rivers Press, 2006.

Bronfenbrenner, Urie. *The Ecology of Human Development: Experiments by Nature and Design*. Cambridge, MA: Harvard University Press, 1979.

Bronfenbrenner, Urie, and Stephen J. Ceci. "Nature-Nurture Reconceptualized in Developmental Perspective: A Bioecological View." *Psychological Review* 101, no. 4 (1994): 568-586.

Bronfenbrenner Urie. "Developmental Ecology Through Space and Time: A Future Perspective." In *Examining Lives in Context: Perspectives on the Ecology of Human Development*, edited by Phyllis Moen, Glen H. Elder, Jr., and Kurt Lüscher, 619-647. Washington, D.C.: American Psychological Association, 1995.

Bronfenbrenner Urie, ed. *Making Human Beings Human: Bioecological Perspectives on Human Development*. Thousand Oaks, CA: Sage Publications, 2005.

Brown, Kenneth G. *Influence: Mastering Life's Most Powerful Skill*. Course Guidebook. Chantilly, VA: The Great Courses, 2013.

Burgoon, Erin M., Marlone D. Henderson, and Arthur B. Markman. "There Are Many Ways to See the Forest for the Trees: A Tour Guide for Abstraction." *Perspectives on Psychological Science* 8, no. 5 (2013): 501-520.

Burke, Peter J. "Perceptions of Leadership in Groups: An Empirical Test of Identity Control Theory." In *Purpose, Meaning, and Action: Control Systems Theories in Sociology*, edited by Kent A. McClelland and Thomas J. Fararo, 267-91. New York: Palgrave Macmillan, 2006.

Burke, Peter J., and Jan E. Stets. *Identity Theory*. Oxford: Oxford University Press, 2009.

Burns, David. *Feeling Good: The New Mood Therapy*. New York: New American Library, 1980.

Butland, Bryony, Susan Jebb, Peter Kopelman, Klim McPherson, Sandy Thomas, Jane Mardell, and Vivienne Parry. *Tackling Obesities: Future Choices – Project Report*. UK: Foresight Programme, Government Office for Science, Department of Innovation Universities and Skills, 2007.

Cabanac, Michael. "Pleasure: The Common Currency." *Journal of Theoretical Biology* 155 (1992): 173-200.

Campion, Michael A., and Robert G. Lord. "A Control Systems Conceptualization of the Goal-Setting and Changing Process." *Organizational Behavior and Human Performance* 30, no. 2 (October 1982): 265-287.

Cannon, Walter B. *The Wisdom of the Body*. New York: W. W. Norton, 1932.

Carey, Timothy A. *The Method of Levels: How to Do Psychotherapy Without Getting in the Way*. Hayward, CA: Living Control Systems Publishing, 2006.

Carey, Timothy A. "Exposure and Reorganization: The What and How of Effective Psychotherapy." *Clinical Psychology Review* 31 (2011): 236-248.

Carey, Timothy A., Warren Mansell, and Sara J. Tai. "A Biopsychosocial Model Based on Negative Feedback and Control." *Frontiers in Human Neuroscience* 8 (February 28, 2014): 1-10.

Carey, Timothy A., Warren Mansell, and Sara J. Tai. *Principles-Based Counselling and Psychotherapy: A Method of Levels Approach*. London: Routledge, 2015.

Carey, Timothy A., and Richard J. Mullan."Evaluating the Method of Levels." *Counselling Psychology Quarterly* 21, no. 3 (September 2008): 247-256.

Carnevale, Anthony P., Leila J. Gainer, and Janice Villet. *Training in America: The Organization and Strategic Role of Training*. San Francisco: Jossey-Bass, 1990.

Carver, Charles S. "Self-Regulation of Action and Affect." In *Handbook of Self-Regulation: Research, Theory, and Applications,* edited by Roy F. Baumeister and Kathleen D. Vohs, 13-39. New York: Guilford Press, 2004.

Carver, Charles S., and Michael F. Scheier. *Attention and Self-Regulation: A Control-Theory Approach to Human Behavior*. New York: Springer-Verlag, 1981.

Carver, Charles S., and Michael F. Scheier. "Control Theory: A Useful Conceptual Framework for Personality-Social, Clinical, and Health Psychology." *Psychological Bulletin* 92, no. 1 (1982): 111-135.

Carver, Charles S., and Michael F. Scheier. *On the Self-Regulation of Behavior*. Cambridge: Cambridge University Press, 1998.

Carver, Charles S., and Michael F. Scheier. "Self-Regulation of Action and Affect." In *Handbook of Self-Regulation: Research, Theory, and Applications.* 2nd ed., edited by Kathleen D. Vohs and Roy F. Baumeister, 3-21. New York: Guilford Press, 2011.

Carver, Charles S., and Michael F. Scheier. "Cybernetic Control Processes and the Self-Regulation of Behavior." In *The Oxford Handbook of Human Motivation*, edited by Richard M. Ryan, 28-42. Oxford: Oxford University Press, 2012.

Chartrand, Tanya L., and John A. Bargh. "The Chameleon Effect: The Perception-Behavior Link and Social Interaction." *Journal of Personality and Social Psychology* 76, no. 6 (June 1999): 893-910.

Chriss, James J. *Social Control: An Introduction*. Cambridge, UK: Polity Press, 2007.

Cialdini, Robert B. *Influence: Science and Practice*. 5th ed. Boston: Pearson, 2009.

Clark, Andy. *Being There: Putting Brain, Body, and World Together Again*. Cambridge, MA: The MIT Press, 1997.

Clark, Herbert H. *Using Language*. Cambridge: Cambridge University Press, 1996.

Clum, George A. *Coping with Panic: A Drug-Free Approach to Dealing with Anxiety Attacks*. Pacific Grove, CA: Thomas Brooks / Cole, 1990.

Cohen, I. Bernard. "Foreword" to *An Introduction to the Study of Experimental Medicine*, by Claude Bernard. Translated by Henry Copley Greene. New York: Dover Publications, 1957. Reprint of 1927 publication by Macmillan & Company.

Cohen, Jacob, and Patricia Cohen. *Applied Multiple Regression/Correlation Analysis for the Behavioral Sciences*. Hillsdale, NJ: Lawrence Erlbaum, 1975.

Cohen, Jacob, Patricia Cohen, Stephen G. West, and Leona S. Aiken. *Applied Multiple Regression/Correlation Analysis for the Behavioral Sciences*. 3rd ed. Mahwah, NJ: Lawrence Erlbaum, 2003.

Coleman, Andrew M. *A Dictionary of Psychology*. 4th ed. Oxford: Oxford University Press, 2015.

Committee on Opportunities in Basic Research in the Behavioral and Social Sciences for the U.S. Military. "Nonverbal Communication." In *Human Behavior in Military Contexts*, edited by James J. Blascovich and Christine R. Hartel , 46-54. Washington, DC: National Academic Press, 2008.

Conrad, Peter, and Joseph W. Schneider. *Deviance and Medicalization: From Badness to Sickness*. ed. Philadelphia: Temple University Press, 1992.

Cummings, William. "Squat Toilets and Cultural Commensurability: Two Texts, Plus Three Photographs I Forgot to Take." *Journal of Mundane Behavior* 1, no. 3 (October 2000): 12 pages, accessed February 11, 2009, http://mundanebehavior.org/issues/v1n3/cummings.htm.

Currie, Shawn R., and Keith G. Wilson. *:60 Second Sleep Ease: Quick Tips to Get a Good Night's Rest*. Far Hills, NJ: New Horizon, 2002.

Cziko, Gary. *Without Miracles: Universal Selection Theory and the Second Darwinian Revolution*. Cambridge, MA: MIT Press, 1995.

Cziko, Gary. *The Things We Do: Using the Lessons of Bernard and Darwin to Understand the What, How, and Why of Our Behavior*. Cambridge, MA: MIT Press, 2000.

Damasio, Antonio. *Descartes' Error: Emotion, Reason, and the Human Brain*. New York: Penguin Books, 1994.

Damasio, Antonio. *The Feeling of What Happens: Body and Emotion in the Making of Consciousness*. San Diego: Harcourt, 1999.

Damasio, Antonio. *Looking for Spinoza: Joy, Sorrow, and the Feeling Brain*. Orlando, FL: Harcourt, 2003.

Dawes, Robyn M. *House of Cards: Psychology and Psychotherapy Built on Myth*. New York: Free Press, 1994.

Dawkins, Richard. *River Out of Eden: A Darwinian View of Life*. New York: Basic Books, 1995.

De Bary, Wm. Theodore, and Irene Bloom, eds. *Sources of Chinese Tradition*. vol. 1, 2nd ed. *From Earliest Times to 1600*. New York: Columbia University Press, 1999.

Dewey, John. *Human Nature and Conduct: An Introduction to Social Psychology*. New York: Henry Holt, 1922/1930. Reprinted by Random House, n.d.

Dijksterhuis, Ap. "Automatic Social Influence: The Perception-Behavior Links as an Explanatory Mechanism for Behavior Matching." In *Social Influence: Direct and Indirect Processes*, edited by Joseph P. Forgas and Kipling D. Williams, 95-108. Philadelphia: Psychology Press, 2001.

Dijksterhuis, Ap. "Why We Are Social Animals: The High Road to Imitation as Social Glue." In *Perspectives on Imitation: From Neuroscience to Social Science*, edited by Susan Hurley and Nick Chater, 2 vols., 207-220. Cambridge, MA: MIT Press, 2005.

Dijksterhuis, Ap, and Pamela K. Smith. "The Unconscious Consumer: Effects of Environment on Consumer Behavior." *Journal of Consumer Psychology* 15, no. 3 (2005): 193-202.

Dombrowski, Stephen U., Falko F. Sniehotta, Allison Avenell, Marie Johnston, Graeme MacLennan, Vera Araújo-Soares. "Identifying Active Ingredients in Complex Behavioural Interventions for Obese Adults with Obesity-Related Co-Morbidities or Additional Risk Factors for Co-Morbidities: A Systematic Review." *Health Psychology Review* 6, no. 1 (2012): 7-32.

Dretske, Fred. *Explaining Behavior: Reasons in a World of Causes*. Cambridge, MA: MIT Press, 1988.

Dreyfus, Hubert L. *Skillful Coping: Essays on the Phenomenology of Everyday Perception and Action*. Edited by Mark A. Wrathall. Oxford: Oxford University Press, 2014.

Dziegielewski, Sophia F., and John S. Wodarski. "Macro-Level Variables as Factors in Human Growth and Development." In *Human Behavior and the Social Environment: Integrating Theory and Evidence-Based Practice*, edited by John S. Wodarski and Sophia F. Dziegielewski, 249-269. New York: Springer, 2002.

Earle, Richard. *The Art of Cause Marketing: How to Use Advertising to Change Personal Behavior and Public Policy*. Lincolnwood, IL: NTC Business Books, 2000.

Ehrlich, Paul R. *Human Natures: Genes, Cultures, and the Human Prospect*. Washington, D.C.: Island Press, 2000.

Elder, Glen H. Jr. *Children of the Great Depression: Social Change in Life Experience*. Boulder, CO: Westview Press, 1999. First published by the University of Chicago Press, 1974.

Elder, Glen H. Jr. "The Life Course Paradigm: Social Change and Individual Development." In *Examining Lives in Context: Perspectives on the Ecology of Human Development*, edited by Phyllis Moen, Glen H. Elder, Jr., and Kurt Lüscher, 101-139. Washington, D.C.: American Psychological Association, 1995.

Eliot, Lise. *What's Going On In There?: How the Brain and Mind Develop in the First Five Years of Life*. New York: Bantam Books, 1999.

Ellis, Albert, and Robert A. Harper. *A New Guide to Rational Living*. North Hollywood, CA: Wilshire Books, 1975.

Ellul, Jacques. *Propaganda: The Formation of Men's Attitudes*. New York: Vintage Books, 1965.

Elsbach, Kimberly D. *Organizational Perception Management*. Mahwah, NJ: Lawrence Erlbaum, 2006.

Elster, Jon. *The Cement of Society: A Study of Social Order*. Cambridge: Cambridge University Press, 1989.

Elster, Jon. *Explaining Social Behavior: More Nuts and Bolts for the Social Sciences*. Cambridge: Cambridge University Press, 2007.

Evans, Gary W. "Child Development and the Physical Environment." *Annual Review of Psychology* 57 (2006): 423-451.

Fairburn, Christopher G. *Overcoming Binge Eating*. New York: Guilford, 1995; 2nd ed., 2013.

Febbraro, Greg A. R., and George A. Clum. "Meta-Analytic Investigation of Self-Regulatory Components in the Treatment of Adult Problem Behaviors." *Clinical Psychology Review* 18, no. 2 (1998): 143-161.

Fiese, Barbara H., Thomas J. Tomcho, Michael Douglas, Kimberly Josephs, Scott Poltrock, and Tim Baker. "A Review of 50 Years of Research on Naturally Occurring Family Routines and Rituals: Cause for Celebration?." *Journal of Family Psychology* 16, no. 4 (2002): 381-390.

Flint, Jonathan, Ralph J. Greenspan, and Kenneth S. Kendler. *How Genes Influence Behavior*. Oxford: Oxford University Press, 2010.

Fogg, B. J., and Dean Eckles. *Mobile Persuasion: 20 Perspectives on the Future of Behavior Change*. Stanford, CA: Stanford Captology Media, 2007.

Fogg, B. J. *Black Path Behavior Guide, Version 0.40*. n.p.: Stanford Persuasive Tech Lab, December 2010.

Fogg, B. J. "Triggers Tell People to 'Do It Now'!." Accessed March 8, 2013 from www.behaviormodel.org/triggers.html.

Ford, Donald H. *Humans as Self-Constructing Living Systems: A Developmental Perspective on Behavior and Personality*. Hillsdale, NJ: Lawrence Erlbaum, 1987.

Ford, Donald H., and Richard M. Lerner. *Developmental Systems Theory: An Integrative Approach*. Newbury Park, CA: SAGE, 1992.

Ford, Edward E. *Love Guaranteed: A Better Marriage in Eight Weeks*. San Francisco: Harper and Row, 1987.

Forgas, Joseph P. *Social Episodes: The Study of Interaction Routines*. London: Academic Press, 1979.

Forgas, Joseph P. "The Affect Infusion Model (AIM): An Integrative Theory of Mood Effects on Cognition and Judgment." In *Theories of Mood and Cognition: A User's Handbook*, edited by Leonard L. Martin and Gerald L. Clore, 99-134. Mahwah, NJ: Lawrence Erlbaum, 2001.

Forgas, Joseph P., and Gordon H. Bower. "Mood Effects on Person-Perception Judgments." *Journal of Personality and Social Psychology* 53, no. 1 (1987): 53-60.

Forssell, Dag, ed. *Perceptual Control Theory: Science and Applications—A Book of Readings*. Hayward, CA: Living Control Systems, 2015.

Frese, Michael, and Doris Fay. "Personal Initiative: An Active Performance Concept for Work in the 21st Century." In *Research in Organizational Behavior*, vol. 23, edited by B. M. Staw and R. M. Sutton, 133-187. Amsterdam: Elsevier Science, 2001.

Frese, Michael, and Dieter Zapf. "Action as the Core of Work Psychology: A German Approach." In *Handbook of Industrial and Organizational Psychology,* vol. 4, edited by Harry C. Triandis, Marvin D. Dunnette, and Leaetta M. Hough, 271-340. Palo Alto, CA: Consulting Psychologists Press, 1994.

Gailliot, Matthew T., Roy F. Baumeister, C. Nathan DeWall, Jon K. Maner, E. Ashby Plant, Dianne M. Tice, Lauren E. Brewer, and Brandon J. Schmeichel. "Self-Control Relies on Glucose as a Limited Energy Source: Willpower Is More Than a Metaphor." *Journal of Personality and Social Psychology* 92, no. 2 (2007): 325-336.

Gailliot, Matthew T., B. Michelle Peruche, E. Ashby Plant, and Roy F. Baumeister. "Stereotypes and Prejudice in the Blood: Sucrose Drinks Reduce Prejudice and Stereotyping." *Journal of Experimental Social Psychology* 45 (2009): 288-300.

Gailliot, Matthew T., E. Ashby Plant, David A. Butz, and Roy F. Baumeister. "Increasing Self-Regulatory Strength Can Reduce the Depleting Effect of Suppressing Stereotypes." *Personality and Social Psychology Bulletin* 33, no. 2 (2007): 281-294.

Gallagher, Winifred. *The Power of Place: How Our Surroundings Shape Our Thoughts, Emotions, and Actions.* New York: Poseidon Press, 1993.

Gamow, David. *Freedom from Stress: How to Take Control of Your Life.* Centennial, CO: Glenbridge, 2006.

Gatz, Margaret, Michael A. Smyer, and Deborah A. DiGilio. "Psychology's Contribution to the Well-Being of Older Americans." *American Psychologist* 74, no.4 (2016): 257-267.

Gibson, James J. *The Ecological Approach to Visual Perception.* Hillsdale, NJ: Lawrence Erlbaum, 1986.

Glass, Thomas A., and Matthew J. McAtee. "Behavioral Science at the Crossroads in Public Health: Extending Horizons, Envisioning the Future." *Social Science & Medicine* 62 (2006): 1650-1671.

Glasser, Robert. "The Reemergence of Learning Theory within Instructional Research." *American Psychologist* 45, no. 1 (January 1990): 29-39.

Goldacre, Ben. *Bad Science.* London: Fourth Estate, 2009.

Gollwitzer, Peter M. "Implementation Intentions: Strong Effects of Simple Plans." *American Psychologist* 54, no. 7 (July 1999): 493-503.

Gollwitzer, Peter M., Ute C. Bayer, and Kathleen C. McCulloch. "The Control of the Unwanted." In *The New Unconscious,* edited by Ran R. Hassin, James S. Uleman, and John A. Bargh, 485-515. Oxford: Oxford University Press, 2005.

Gollwitzer, Peter M., and Gabriele Oettingen. "Planning Promotes Goal Striving." In *Handbook of Self-Regulation: Research, Theory, and Applications,* 2nd ed., edited by Kathleen D. Vohs and Roy F. Baumeister, 162-185. New York: Guilford Press, 2011.

Gorman, Nicholas, Jeffery A. Lackney, Kimberly Rollings, and Terry T.-K. Huang. "Designer Schools: The Role of School Space and Architecture in Obesity Prevention." *Obesity* 15, no. 11 (November 2007): 2521-2530.

Gregory, Richard. *Seeing Through Illusions*. Oxford: Oxford University Press, 2009.

Grilly, David M., and John D. Salamone. *Drugs, Brain, and Behavior*. 6th ed. Boston: Pearson, 2012.

Hadnagy, Christopher. *Social Engineering: The Art of Human Hacking*. Indianapolis: Wiley, 2011.

Hamilton, Virginia. *In the Beginning: Creation Stories from Around the World*. San Diego: Harcourt Brace Jovanovich, 1988.

Hampson, Elizabeth. "Sex Differences in Human Brain and Cognition: The Influence of Sex Steroids in Early and Adult Life." In *Behavioral Endocrinology,* 2nd ed., edited by Jill B. Becker, S. Marc Breedlove, David Crews, and Margaret M. McCarthy, 579-628. Cambridge, MA: MIT Press, 2002.

Harris, Jennifer L., John A. Bargh, and Kelly D. Brownell. "Priming Effects of Television Food Advertising on Eating Behavior." *Health Psychology* 28, no. 4 (2009): 404-13.

Harris, Richard Jackson. *A Cognitive Psychology of Mass Communication*. 5th ed. New York: Routledge, 2009.

Harris, Richard Jackson, and Fred W. Sanborn. *A Cognitive Psychology of Mass Communication*. 6th ed. New York: Routledge, 2014.

Hayakawa, S. I., and Alan R. Hayakawa. *Language in Thought and* Action. 5th ed. Orlando, FL: Harvest, 1992.

Heath, Chip, and Dan Heath. *SWITCH: How to Change Things When Change is Hard*. New York: Broadway Books, 2010.

Heiman, Julia R., and Joseph LoPiccola. *Becoming Orgasmic: A Sexual and Personal Growth Program for Women*. New York: Prentice Hall, 1976. Reprinted 1988.

Heise, David R. *Causal Analysis*. New York: Wiley, 1975.

Heise, David R. *Expressive Order: Confirming Sentiments in Social Actions*. New York: Springer, 2007.

Herbig, Katherine L., and Martin F. Wiskoff. *Espionage Against the United States by American Citizens 1947-2001*. Technical Report 02-5. Monterey, CA: Defense Personnel Security Research Center, July 2002.

Herzog, Hal. "Playing Dead as a Defense Strategy Against Angry Bears and Rapists." *Psychology Today,* July 29, 2010, www.psychologytoday.com/blog/animals-and-us/201007/playing-dead-defense-against-angry-bears-and-rapists.

Houk, James C. "Homeostasis and Control Principles." In *Medical Physiology,* 14th ed., edited by Vernon B. Mountcastle, 246-267. St. Louis: C. V. Mosby, 1980.

Howard, Pierce J. *The Owner's Manual for the Brain.* Austin: Bard Press, 1994.

Huang, Julie Y., and John A. Bargh. "The Selfish Goal: Autonomously Operating Motivational Structures as the Proximate Cause of Human Judgment and Behavior." *Behavioral and Brain Science* 37 (2014): 154, 162-4.

Huesmann, L. Rowell. "Imitation and the Effects of Observing Media Violence on Behavior." In *Perspectives on Imitation,* vol.2, eds. Susan Hurley and Nick Chater, 257-66. Cambridge, MA: MIT Press, 2005.

Hyland, M. E. "Control Theory and Psychology: A Tool for Integration and a Heuristic for New Theory." In *Volitional Action: Conation and Control,* edited by W. A. Hershberger, 353-369. Amsterdam: North-Holland, 1989.

Ingold, Tim. *The Perception of the Environment: Essays on Livelihood, Dwelling and Skill.* London: Routledge, 2000.

Jaccard, James, and Jacob Jacoby. *Theory Construction and Model-Building Skills: A Practical Guide for Social Scientists.* New York: Guilford Press, 2010.

Jahoda, Gustav. "Critical Comments on Experimental, Discursive, and General Social Psychology." *Journal for the Theory of Social Behaviour* 43, no. 3 (2012): 341-360.

James, William. *The Principles of Psychology*. 2 vols. New York: Cosimo, 2007. First published 1890.

Jarvis, Peter. *Toward a Comprehensive Theory of Human Learning: Lifelong Learning and the Learning Society,* vol. 1. London: Routledge, 2006.

Johnson, Kenneth G. *General Semantics: An Outline Survey*. 3rd ed. rev. Fort Worth: Institute of General Semantics, 2004.

Johnson, Russell E., Chu-Hsiang Chang, and Robert G. Lord. "Moving from Cognition to Behavior: What the Research Says." *Psychological Bulletin* 132, no. 3 (2006): 381-415.

Johnston, Victor S. *Why We Feel: The Science of Human Emotions*. n.p.: Helix Books, 1999.

Joye, Yannick. "Architectural Lessons from Environmental Psychology: The Case of Biophilic Architecture." *Review of General Psychology* 11, no. 4 (2007): 305-328.

Kahneman, Daniel. *Thinking Fast and Slow*. New York: Farrar, Straus and Giroux, 2011.

Kandel, Eric R. *In Search of Memory*. New York: W. W. Norton, 2006.

Kandel, Eric R., James H. Schwartz, and Thomas M. Jessell, eds. *Principles of Neural Science*. 4th ed. New York: McGraw-Hill, 2000.

Kandel, Eric R., James H. Schwartz, Thomas M. Jessell, Steven A. Siegelbaum, and A. J. Hudspeth, eds. *Principles of Neural Science*. 5th ed. New York: McGraw-Hill, 2013.

Kanizsa, Gaetano. "Margini quasi-percettivi in campi con stimolazione omogenea." *Rivista di Psicologia* 49, no. 1 (1955): 7–30.

Katz, Daniel, and Robert L. Kahn. *The Social Psychology of Organizations*, 2nd ed. New York: John Wiley, 1978.

Keizer, Kees, Siegwart Lindenberg, and Linda Steg. "The Spreading of Disorder." *Science* 322 (December 12, 2008): 1681-1685.

Kelly, Harold H., John G. Holmes, Norbert L. Kerr, Harry T. Reis, Caryl E. Rusbult, and Paul A. M. Van Lange. *An Atlas of Interpersonal Situations*. Cambridge: Cambridge University Press, 2003.

Kirsch, Irving, ed. *How Expectancies Shape Experience*. Washington, D.C.: American Psychological Association, 1999.

Kirsch, Irving, and Steven Jay Lynn. "Automaticity in Clinical Psychology." *American Psychologist* 54, no. 7 (July 1999): 504-515.

Kirst-Ashman, Karen K. *Human Behavior, Communities, Organizations, and Groups in the Macro Social Environment*. Belmont, CA: Brooks/Cole, 2008.

Klein, Howard. "An Integrated Control Theory Model of Work Motivation." *Academy of Management Review* 14, no. 2 (1989): 150-172.

Klein, Howard J. "Control Theory and Understanding Motivated Behavior: A Different Conclusion." *Motivation and Emotion* 15, no. 1 (1991): 29-44.

Klinedinst, Lloyd. "My Reply to Richard and Rick and Martin and Warren." Attachment to a memo titled "Habit-my revised working definition," sent July 5, 2015 by e-mail to csgnet@lists.illinois.edu.

Klöckner, Christian A., Ellen Matthies, and Marcel Hunecke. "Problems of Operationalizing Habits and Integrating Habits in Normative Decision-Making Models." *Journal of Applied Social Psychology* 33, no. 2 (February 2003): 397.

Knudsen, Eric I. "Early Experience and Sensitive Periods." In Squire et al., *Fundamental Neuroscience*, 517-32.

Korzybski, Alfred. *Science and Sanity: An Introduction to Non-Aristotelian Systems and General Semantics*. 5th ed. Fort Worth: Institute of General Semantics, 1994.

Korzybski, Alfred. *Selections from Science and Sanity: An Introduction to Non-Aristotelian Systems and General Semantics*. 2nd ed. Fort Worth: Institute of General Semantics, 2010.

Kotler, Philip, and Nancy R. Lee. *Social Marketing: Influencing Behaviors for Good*. 3rd ed. Los Angeles: SAGE, 2008.

Kram, Kathy E. *Mentoring at Work*. Glenview, IL: Scott, Foresman, 1985.

Lakoff, George, *Don't Think of an Elephant!: Know Your Values and Frame the Debate*, White River Junction, VT: Chelsea Publishing, 2004.

Lama, Dalai, and Howard C. Cutler. *The Art of Happiness: A Handbook for Living*. New York: Riverhead Books, 1998.

Langer, Ellen J. *Mindfulness*. Reading, MA: Addison-Wesley, 1989.

Laszlo, Ervin. *The Systems View of the World: A Holistic Vision for Our Time*. Cresskill, NJ: Hampton Press, 1996.

Lawrence, Paul R., and Nitin Nohria. *Driven: How Human Nature Shapes Our Choices*. San Francisco: Jossey-Bass, 2002.

Le Cheminant, Wayne, and John M. Parrish, eds. *Manipulating Democracy: Democratic Theory, Political Psychology, and Mass Media*. New York: Routledge, 2011.

Leeming, David A. *Creation Myths of the World: An Encyclopedia*. 2nd ed. Santa Barbara: ABC-CLIO, 2010.

Leutzinger, Joseph, and John Harris. *Why and How People Change Health Behaviors: A Collection of Success Stories*, vol. 1. Omaha: Health Improvement Solutions, 2005.

Levitis, Daniel A., William Z. Lidicker, Jr., and Glenn Freund. "Behavioural Biologists Do Not Agree on What Constitutes Behaviour." *Animal Behaviour* 78, no. 1 (July 2009): 103-110.

Lewinsohn, Peter M., Ricardo F. Munoz, Mary Ann Youngren, and Antonette M. Zeiss. *Control Your Depression*. New York: Simon and Schuster, 1986. Revised edition 1992.

Liederbach, John. "'Pass the Trash': The Mortgage Default Crisis as State-Corporate Crime." In *Social Control: Informal, Legal and Medical*, edited by James J. Chriss, 17-41. Bingley, UK: Emerald Group Publishing, 2010.

Linton, Ralph. *The Study of Man: An Introduction*. n.p.: Appleton-Century-Crofts, 1936.

Lipton, Peter. "Causation and Explanation." In *The Oxford Handbook of Causation*, edited by Helen Beebe, Christopher Hitchcock, and Peter Menzies, 619-31. Oxford: Oxford University Press, 2009.

Locke, Edwin A., and Gary P. Latham. *Goal Setting: A Motivational Technique That Works!* Englewood Cliffs: Prentice-Hall, 1984.

Locke, Edwin A., and Gary P. Latham. *A Theory of Goal Setting and Task Performance*. Englewood Cliffs: Prentice-Hall, 1990.

Lord, Robert G., and Paul E. Levy. "Moving from Cognition to Action: A Control Theory Perspective." *Applied Psychology: An International Review* 43, no. 3 (1994): 335-398.

MacKenzie, Meagan, Peter G. Mezo, and Sarah E. Francis. "A Conceptual Framework for Understanding Self-Regulation in Adults." *New Ideas in Psychology* 30, no. 2 (August 2012): 155-165.

Maddux, James E. "Habit, Health, and Happiness." *Journal of Sport and Exercise* 19 (1997): 331-346.

Mager, Robert F., and Peter Pipe. *Analyzing Performance Problems: or You Really Oughta Wanna*. 2nd ed. Belmont, CA: Lake Publishing, 1984.

Mager, Robert F., and Peter Pipe. *Analyzing Performance Problems: or You Really Oughta Wanna*, 3rd ed. Atlanta: Center for Effective Performance, 1997.

Magistretti, Pierre J. "Brain Energy Metabolism." In *Fundamental Neuroscience*, 3rd ed., edited by Larry Squire, Darwin Berg, Floyd Bloom, Sascha du Lac, Anirvan Ghosh, and Nicholas Spitzer, 271-293. Amsterdam: Elsevier, 2008.

Magistretti, Pierre J. "Low-Cost Travel in Neurons." *Science* 325 (September 11, 2009): 1349-51.

Magnusson, David. "Wanted: A Psychology of Situations." In *Toward a Psychology of Situations: An Interactional Perspective*, edited by David Magnusson, 9-32. Hillsdale, NJ: Lawrence Erlbaum, 1981.

Mansell, Warren. "Control Theory and Psychopathology: An Integrative Approach." *Psychology and Psychotherapy: Theory, Research and Practice* 78 (2005): 141-178.

Mansell, Warren, and Timothy A. Carey. "A Century of Psychology and Psychotherapy: Is an Understanding of 'Control' the Missing Link Between Theory, Research, and Practice." *Psychology and Psychotherapy: Theory, Research and Practice* 82, no. 3 (2009): 337-353.

Manz, Charles C., and Henry P. Sims, Jr. "Vicarious Learning: The Influence of Modeling on Organizational Behavior." *Academy of Management Review* 6, no. 1 (January 1981): 105-113.

Marcus, Gary. "Making the Mind: Why We've Misunderstood the Nature-Nurture Debate." In *Taking Sides: Clashing Views on Psychological Issues,* 15th ed., edited by Brent Slife, 135-144. Boston, McGraw-Hill, 2008.

Marken, Richard S. "Perceptual Organization of Behavior: A Hierarchical Control Model of Coordinated Action." *Journal of Experimental Psychology: Human Perception and Performance* 12, no. 3 (1986): 267-276, reprinted in Marken, *Mind Readings,* 159-184.

Marken, Richard S. "The Nature of Behavior: Control as Fact and Theory." *Behavioral Science* 33, no. 3 (July 1988): 196-206

Marken, Richard S. *Mind Readings: Experimental Studies of Purpose.* Durango, CO: Control Systems Group, 1992.

Marken, Richard. "Controlled Variables: Psychology as the Center Fielder Views It." *American Journal of Psychology* 114, no. 2 (Summer 2001): 259-281.

Marken, Richard. "Looking at Behavior Through Control Theory Glasses." *Review of General Psychology* 6, no. 3 (2002a): 260-270.

Marken, Richard S. *More Mind Readings: Methods and Models in the Study of Purpose.* St. Louis: Newview, 2002b.

Marken, Richard S. "You Say You Had a Revolution: Methodological Foundations of Closed-Loop Psychology." *Review of General Psychology* 13, no. 2 (2009): 137-145.

Marken, Richard S. "Making Inferences About Intention: Perceptual Control Theory as a 'Theory of Mind' for Psychologists." *Psychological Reports* 113, no. 1 (2013): 1-18.

Marken, Richard S. *Doing Research on Purpose: A Control Theory Approach to Experimental Psychology*. St. Louis: New View, 2014.

Marken, Richard S., and Timothy A. Carey. *Controlling People: The Paradoxical Nature of Being Human*. Samford Valley, Australia: Australian Academic Press, 2015.

Marken, Richard S., and Warren Mansell. "Perceptual Control as a Unifying Concept in Psychology" *Review of General Psychology* 17, no. 2 (2013): 190-195.

Markway, Barbara G., Cheryl N. Carmin, C. Alec Pollard, and Teresa Flynn. *Dying of Embarrassment: Help for Social Anxiety and Phobia.* Oakland, CA: New Harbinger, 1992.

Martin, Paul. *Counting Sheep: The Science and Pleasures of Sleep and Dreams*. New York: Thomas Dunne Books, 2002.

Maslow, Abraham. *Motivation and Personality*. New York: Harper, 1954.

Maturana, Humberto R., and Francisco J. Varela. *Autopoiesis and Cognition: The Realization of the Living*. Dordrecht, Holland: D. Reidel Publishing Company, 1980.

Maturana, Humberto R., and Francisco J. Varela. *The Tree of Knowledge: The Biological Roots of Human Understanding*. rev. ed. Boston: Shambhala, 1992.

McClelland, Kent. "The Collective Control of Perceptions: Toward a Person-Centered Sociology." Revised draft of a paper presented to the Annual Meeting of the Control Systems Group, Flagstaff, Arizona, 1996. Accessed September 26, 2005.

McClelland, Kent. "The Collective Control of Perceptions: Constructing Order from Conflict." *International Journal of Human-Computer Studies* 60 (2004): 65-99.

McClelland, Kent. "Understanding Collective Control Processes." In *Purpose, Meaning, and Action* edited by Kent A. McClelland and Thomas J. Fararo, 31-56. New York: Palgrave Macmillan, 2006.

McClintok, Martha K. "Menstrual Synchrony and Suppression." *Nature* 229 (January 22, 1971): 244-245.

McCormick, David A., and Gary L. Westbrook. "Sleep and Dreaming." In *Principles of Neural Science*. 5th ed., edited by Eric R. Kandel, James H. Schwartz, Thomas M. Jessell, Steven A. Siegelbaum, and A. J. Hudspeth, 1140-1158. New York: McGraw-Hill, 2013.

McCrea, Michael A. *Mild Traumatic Brain Injury and Postconcussion Syndrome: The New Evidence Base for Diagnosis and Treatment.* Oxford: Oxford University Press, 2008.

McLaughlin, Katie A., Margaret A. Sheridan, Florin Tibu, Nathan A. Fox, Charles H. Zeanah, and Charles A. Nelson. "Causal Effects of the Early Caregiving Environment on Development of Stress Response Systems in Children." *Proceedings of the National Academy of Sciences* (April 20, 2015); 201423363 DOI: 10.1073/npas.1423363112.

McNulty, James K., Michael A. Olsen, Andrea L. Meltzer, and Matthew J. Shaffer. "Though They May Be Unaware, Newlyweds Implicitly Know Whether Their Marriage Will Be Satisfying." *Science* 342 (November 29, 2013): 1119-1120.

McPhail, Clark. *The Myth of the Madding Crowd*. New York: Aldine de Gruyter, 1991.

Meadows, Donella H. *Thinking in Systems: A Primer*. White River Junction, VT: Chelsea Green Publishing, 2008.

Michie, Susan, Charles Abraham, Craig Whittington, John McAteer, and Sunjai Gupta. "Effective Techniques in Healthy Eating and Physical Activity Interventions: A Meta-Regression." *Health Psychology* 28, no. 6 (2009): 690-701.

Michie, Susan, Stefanie Ashford, Falko F. Sniehotta, Stephen U. Dombrowski, Alex Bishop, & David P. French. "A Refined Taxonomy of Behaviour Change Techniques to Help People Change Their Physical Activity and Healthy Eating Behaviours: The CALO-RE Taxonomy." *Psychology and Health* 26, no. 11 (November 2011): 1479-1498.

Michie, Susan, Nichola Rumsey, Anna Fussell, Wendy Hardeman, Marie Johnston, Stanton Newman, and Lucy Yardley. *Improving Health: Changing Behaviour: NHS Health Trainer Handbook*. London: Department of Health/British Psychological Society, 2008.

Miller, Joan G. "Culture and the Development of Everyday Social Explanation." *Journal of Personality and Social Psychology* 46, no. 5 (1984): 691-978.

Miller, William R., and Stephen Rollnick. *Motivational Interviewing: Helping People Change*. 3rd ed. New York: Guilford Press, 2013.

Milliman, Ronald E. "Using Background Music to Affect the Behavior of Supermarket Shoppers." *Journal of Marketing* 46, no. 3 (Summer 1982): 86-91.

Mingers, John. *Self-Producing Systems: Implications and Applications of Autopoiesis*. New York: Plenum Press, 1995.

Mish, Frederick C., ed. *Webster's Ninth New Collegiate Dictionary*. Springfield: Merriam-Webster, 1991.

Moen, Phyllis, Glen H. Elder, Jr., and Kurt Lüscher, eds. *Examining Lives in Context: Perspectives on the Ecology of Human Development*. Washington, D.C.: American Psychological Association, 1995.

Mohr, Lawrence B. *The Causes of Human Behavior: Implications for Theory and Method in the Social Sciences*. Ann Arbor: University of Michigan Press, 1996.

Moos, Rudolf H. *The Human Context: Environmental Determinants of Behavior*. Malabar, FL: Robert E. Krieger, 1986. First published by John Wiley, 1976.

Morin, Charles M. *Relief from Insomnia*. New York: Main Street Books/Doubleday, 1996.

Morris, John. *The Young Earth: The Real History of the Earth—Past, Present, and Future*. Green Forest, AR: Master Books, 2007.

Morris, Lydia, Warren Mansell, and Phil McEvoy. "The Take Control Course: Conceptual Rationale for the Development of a Transdiagnostic Group for Common Mental

Problems." *Frontiers in Psychology* 7, Article 99 (February 2016): doi: 10.3389/fpsyg.2016.00099.

Morris, Michael W., and Kaiping Peng. "Culture and Cause: American and Chinese Attributions for Social and Physical Events." *Journal of Personality and Social Psychology* 67, no. 6 (1994): 949-971.

Mrosovsky, Nicholas. *Rheostasis: The Physiology of Change*. Oxford: Oxford University Press, 1990.

Munz, Peter. *Critique of Impure Reason: An Essay on Neurons, Somatic Markers, and Consciousness*. Westport, CT: Praeger, 1999.

Muraven, Mark. "Building Self-Control Strength: Practicing Self-Control Leads to Improved Self-Control Performance." *Journal of Experimental Social Psychology* 46, no. 2 (2010): 465-468.

Muraven, Mark, Roy F. Baumeister, and Dianne M. Tice. "Longitudinal Improvement of Self-Regulation Through Practice: Building Self-Control Strength Through Repeated Exercise." *Journal of Social Psychology* 139, no. 4 (1999): 446-457.

NAFTA at Twenty: Accomplishments, Challenges, and the Way Forward: Hearings of the Subcommittee on the Western Hemisphere. 113 Congress, 2nd session, 15 January 2014 (1), Serial No. 113-112. Washington, DC: U.S. Government Printing Office, 2014.

Nalbone, David P., Kevin P. Lee, Amanda R. Suroviak, and Jennifer M. Lannon. "The Effects of Social Norms on Male Hygiene." *Individual Difference Research* 3, no. 3 (2005): 171-176.

Nathan, Peter E., and Jack M. Gorman, eds. *A Guide to Treatments that Work*. Oxford: Oxford University Press, 2007

National Institute of Mental Health. *Anxiety Disorders*. Washington, DC: National Institute of Mental Health, 2009.

National Institute of Mental Health. *The Teen Brain: Still Under Construction*. Publication no. 11-4929. Bethesda, MD: National Institute of Mental Health, 2011.

National Research Council and Institute of Medicine. *U.S. Health in International Perspective: Shorter Lives, Poorer Health*. Washington, D.C.: National Academies Press, 2013.

Nelson, Randy J. *An Introduction to Behavioral Endocrinology*. 3rd ed. Sunderland, MA: Sinauer Associates, 2005.

Nelson, Randy J. *An Introduction to Behavioral Endocrinology*. 4th ed. Sunderland, MA: Sinauer Associates, 2011.

Nickols, Fred. *Helping People Hit Their Performance Targets*. Mount Vernon, Ohio: Distance Consulting, September 1, 2010.

Nickols, Fred. *Manage Your Own Performance: No One Else Can*. Mount Vernon, Ohio: Distance Consulting, October 4, 2010.

Nickols, Fred. *The Target Model: A Mainly Visual Presentation*. Mount Vernon, Ohio: Distance Consulting, September 6, 2012.

Nickols, Fred. *A Control Theory View of Human Performance in the Workplace*. Mount Vernon, Ohio: Distance Consulting, 2013a.

Nickols, Fred. *Tools for Knowledge Workers: PCT 101: A Primer*. Mount Vernon, Ohio: Distance Consulting, 2013b.

Nisbett, Richard E., and Timothy DeCamp Wilson. "Telling More Than We Can Know: Verbal Reports on Mental Processes." *Psychological Review* 84, no 3 (May 1977): 231-259.

Noë, Alva. *Action in Perception*. Cambridge, MA: MIT Press, 2004.

Norcross, John C. *Changeology: 5 Steps to Realizing Your Goals and Resolutions*. New York: Simon and Schuster, 2012.

Norcross, John C., Linda F. Campbell, John M. Grohol, John W. Santrock, Florin Selagea, and Robert Sommer. *Self-Help That Works: Resources to Improve Emotional Health and Strengthen Relationships*. Oxford: Oxford University Press, 2013.

North, Adrian C., David J. Hargreaves, and Jennifer McKendrick. "In-Store Music Affects Product Choice." *Nature* 390 (November 13, 1997): 132.

North, Adrian C., David J. Hargreaves, and Jennifer McKendrick. "The Influence of In-Store Music on Wine Selections." *Journal of Applied Psychology* 84, no. 2 (1999): 271-276.

Northcutt, Wendy. *The Darwin Awards: Evolution in Action*. New York: Plume, 2000.

Northcutt, Wendy. *The Darwin Awards II: Unnatural Selection*. New York: Plume, 2001.

Northcutt, Wendy. *The Darwin Awards: Survival of the Fittest*. New York: Plume, 2003.

Ogden, Jane. "Some Problems with Social Cognition Models: A Pragmatic and Conceptual Analysis." *Health Psychology* 22, no. 4 (July 2003): 424-428.

Orbell, Sheina, Patrick Lidierth, Caroline J. Henderson, Nicolas Geeraert, Claudia Uller, Ayse K. Uskul, and Maria Kyriakaki. "Social-Cognitive Beliefs, Alcohol, and Tobacco Use: A Prospective Community Study of Change Following a Ban on Smoking in Public Places." *Health Psychology* 28, no. 6 (2009): 753-761.

Ouellette, Judith A., and Wendy Wood. "Habit and Intention in Everyday Life: The Multiple Processes by Which Past Behavior Predicts Future Behavior." *Psychological Bulletin* 124, no. 1 (July 1998): 54-74.

Panneton, W. Michael. "The Mammalian Diving Response: An Enigmatic Reflex to Preserve Life?" *Physiology* 28, no. 5 (2013): 284-297.

Pavloski, Raymond P. "The Physiological Stress of Thwarted Intention." In *Volitional Action: Conation and Control,* edited by Wayne A. Hershberger, 215-32. Amsterdam: Holland-North, 1989.

Pearsall, Paul. *The Last Self-Help Book You'll Ever Need.* New York: Basic Books, 2005.

Pease, Allan, and Barbara Pease. *The Definitive Book of Body Language.* New York: Bantam Books, 2004.

Peper, Jiska S., and Ronald E. Dahl. "The Teenage Brain: Surging Hormones—Brain-Behavior Interactions During Puberty." *Psychological Science* 22, no. 2 (2013): 134-139.

Perloff, Richard M. *The Dynamics of Persuasion: Communication and Attitudes in the 21st Century.* Mahwah, NJ: Lawrence Erlbaum, 2003.

Petty, Richard E. "Subtle Influences on Judgment and Behavior: Who is Most Susceptible?." In *Social Influence: Direct and Indirect Processes,* edited by Joseph P. Forgas and Kipling D. Williams, 129-146. Philadelphia: Psychology Press, 2001.

Pfau, Richard H. *How to Identify the Training Needs of Employees.* 3rd ed. Mansfield, CT: The Workforce Training Group, 2007.

Pfau, Richard H. "Using Perceptual Control Theory as a Framework for Influencing and Controlling Behavior." Paper prepared for presentation at the Control Systems Group International Conference, Boulder, Colorado, July 17-21, 2012.

Piantadosi, Claude A. *The Biology of Human Survival: Life and Death in Extreme Environments.* Oxford: Oxford University Press, 2003.

Powers, Mary. "Control Theory: A New Direction for Psychology, A Reply to Todd Nelson." In *Perceptual Control Theory: An Overview of the Third Grand Theory in Psychology,* edited by Dag Forssell, 59-61. Hayward, CA: Living Control Systems Publishing, 2016.

Powers, William T. *Behavior: The Control of Perception.* Chicago: Aldine, 1973.

Powers, William T. "Quantitative Analysis of Purposive Systems: Some Spadework at the Foundations of Scientific Psychology." *Psychological Review* 85, no. 5 (September 1978): 417-435. Reprinted in *Living Control Systems: Selected Papers of William*

T. Powers, edited by Gregory Williams, 129-165. New Canaan, CT: Benchmark Publications, 1989.

Powers, William T. "The Nature of Robots. Part 4: Looking for Controlled Variables." *BYTE* 4 (September 1979): 96-118

Powers, William T. "An Outline of Control Theory." In *Conference Workbook for "Texts in Cybernetic Theory": An In-Depth Exploration of the Thought of Humberto Maturana, William T. Powers, Ernst Von Glaserfeld.* Felton, CA: American Society of Cybernetics, October 1988.

Powers, William T. "A Feedback Model for Behavior: Application to a Rat Experiment." In *Living Control Systems: Selected Papers of William T. Powers,* edited by Gregory Williams, 47-59. New Canaan, CT: Benchmark Publications, 1989. Reprinted from *Behavioral Science* 16, no. 6 (November 1971): 558-563.

Powers, William T. "A Cybernetic Model for Research in Human Development," In *Living Control Systems: Selected Papers of William T.* Powers, edited by Gregory Williams, 167-219. New Canaan, CT: Benchmark Publications, 1989. Reprinted from *A Cybernetic Approach to the Assessment of Children.* Edited by Mark N. Ozer, 11-66. Boulder: Westview Press, 1979.

Powers, William T. *Living Control Systems: Selected Papers of William T. Powers.* Gravel Switch: KY: Control Systems Group, 1989. Reprint, New Canaan, CT: Benchmark, 2005.

Powers, William T. *Living Control Systems II: Selected Papers of William, T. Powers.* Gravel Switch: KY: Control System Group, 1992.

Powers, William T. *Making Sense of Behavior: The Meaning of Control.* New Canaan, CT: Benchmark Publications, 2004.

Powers, William T. *Behavior: The Control of Perception.* 2nd ed. New Canaan, CT: Benchmark Publications, 2005.

Powers, William T. *Living Control Systems III: The Fact of Control.* Bloomfield, NJ: Benchmark Publications, 2008.

Powers, W. T. "The World According to PCT." Unpublished paper, 2013, attached to an e-mail by Alice McElhone to csgnet<csgnet@listserve.illinois.edu, December 5, 2013.

Powers, William T., Bruce Abbott, Timothy A. Carey, David M. Goldstein, Warren Mansell, Richard S. Marken, Bruce Nevin, Richard Robertson, and Martin Taylor. *Perceptual Control Theory: A Model for Understanding the Mechanisms and Phenomena of Control.* Published August 2011, and downloaded from http://pctweb.org/PCTUnderstanding.pdf.

Powers, W. T., R. K. Clark, and R. L. McFarland. "A General Feedback Theory of Human Behavior: Part I." *Perceptual and Motor Skills* 11 (1960): 71-88.

Powers, W. T., R. K. Clark, and R. L. McFarland. "A General Feedback Theory of Human Behavior: Part II." *Perceptual and Motor Skills* 11 (1960): 309-323.

Provine, Robert R. *Curious Behavior: Yawning, Laughing, Hiccupping and Beyond.* Cambridge, MA: Belknap Press of Harvard University Press, 2012.

Quinn, Jeffrey M., Anthony Pascoe, Wendy Wood, and David T. Neal. "Can't Control Yourself? Monitor Those Bad Habits." *Personality and Social Psychology Bulletin* 36, no. 4 (2010): 499-511.

Raabe, Babette, Michael Frese, and Terry A. Beehr. "Action Regulation Theory and Career Self-Management." *Journal of Vocational Behavior* 70, no. 2 (April 2007): 297-311.

Radcliff, Benjamin, and Amy Radcliff. *Understanding Zen.* Boston: Charles E. Tuttle, 1993.

Ramnerö, Jonas, and Niklas Törneke. *The ABCs of Human Behavior: Behavioral Principles for the Practicing Clinician.* Oakland, CA: New Harbinger Publications, 2008.

Randall, Allan F. "Living in a Cybernetic Universe: A Constructivist Cybernetics of Perception." Paper presented at the 50th Anniversary Conference of the American Society for Cybernetics, Washington, DC, August 2014.

Random House Dictionary of the English Language. 2nd ed. New York: Random House, 1987.

Reynolds, John H., Jacqueline P. Gottlieb, and Sabine Kastner. "Attention." In *Fundamental Neuroscience,* 3rd ed., edited by Larry Squire, Darwin Berg, Floyd Bloom, Sascha du lac, Anirvan Ghosh, and Nicholas Spitzer, 1113-32. Amsterdam: Elsevier, 2008.

Reynolds, Paul Davidson. *A Primer in Theory Construction.* 1971. Reprinted. Boston: Pearson, 2007.

Rhoads, Kelton. "Everyday Influence." Working Psychology website, accessed October 29, 2015. http://workingpsychology.com/evryinfl.html.

Richardson, George P. *Feedback Thought in Social Science and Systems Theory.* Waltham, MA: Pegasus Communications, 1999.

Riehle, Alexa. "Neuronal Correlates of Context-Related Behavior." *Journal of Pragmatics* 35 (2003): 485-504.

Ries, Al, and Jack Trout. *Positioning: The Battle for Your Mind.* New York: McGraw-Hill, 2001.

Rizzolatti, Giacomo, and Laila Craighero. "The Mirror-Neuron System." *Annual Review of Neuroscience* 27 (2004): 169-92.

Robertson, Richard J., and William T. Powers, eds. *Introduction to Modern Psychology: The Control-Theory View*. Gravel Switch, KY: Control System Group, 1990.

Robinson, Dawn T. "Control Theories in Sociology." *Annual Review of Sociology* 33 (2007): 157-174

Rose, Steven. *Lifelines: Biology Beyond Determinism*. Oxford: Oxford University Press, 1997.

Rosenblueth, Arturo, Norbert Weiner, and Julian Bigelow. "Behavior, Purpose, and Teleology." *Philosophy of Science* 10, no. 1 (January 1943): 18-24.

Ross, Lee, and Richard E. Nisbett. *The Person and the Situation: Perspectives of Social Psychology*. New York: McGraw-Hill, 1991.

Runkel, Philip J. *Casting Nets and Testing Specimens: Two Grand Methods of Psychology*. New York: Praeger, 1990.

Runkel, Philip J. *People as Living Things: The Psychology of Perceptual Control*. Hayward, CA: Living Control Systems Publishing, 2005.

Russell, James A. "Core Affect and the Psychological Construction of Emotion." *Psychological Review* 110, no. 1 (January 2003): 145-172.

Rutter, Michael. *Genes and Behavior: Nature-Nurture Interplay Explained*. Malden, MA: Blackwell Publishing, 2006.

Rutter, Michael, Henri Giller, and Ann Hagell. *Antisocial Behavior by Young People*. Cambridge: Cambridge University Press, 1998.

Sandis, Constantine. *The Things We Do and Why We Do Them*. New York: Palgrave Macmillan, 2012.

Sapolsky, Robert M. *Why Zebras Don't Get Ulcers*. 3rd ed. New York: Henry Holt, 2004.

Sapolsky, Robert. *Biology and Human Behavior: the Neurological Origins of Individuality*. Part 1. 2nd ed. Chantilly, VA: Teaching Company, 2005a.

Sapolsky, Robert. *Biology and Human Behavior: the Neurological Origins of Individuality*. Part 2. 2nd ed. Chantilly, VA: Teaching Company, 2005b.

Scales, Peter C., with Peter L. Benson, Marc Mannes, Nicole R. Hintz, Eugene C. Roehlkepartain, and Theresa K. Sullivan. *Other Peoples Kids: Social Expectations and American Adults' Involvement with Children and Adolescents*. New York: Kluwer Academic/Plenum Publishers, 2003.

Scharfstein, Ben-Ami. *The Dilemma of Context.* New York: New York University Press, 1989.

Schoggen, Phil. *Behavior Settings: A Revision and Extension of Roger G. Barker's Ecological Psychology*. Stanford: Stanford University Press, 1989.

Scott, Eugenie C. *Evolution vs. Creationism: An Introduction*. 2nd ed. Berkeley: University of California Press, 2009.

Seldes, George. *The Great Quotations*. Secaucus, NJ: Castle Books, 1966.

Shabo, Magedah E. *Techniques of Propaganda and Persuasion*. Clayton, DE: Prestwick House, 2008.

Sheeran, Paschal, and Thomas L. Webb. "From Goals to Action." In *Goal-Directed Behavior,* edited by Henk Aarts and Andrew J. Elliot, 175-202. New York: Psychology Press, 2012.

Sherpa, Lhamo Yanghen. "Fight, Flight, or Freeze." *Republica* (January 27, 2013). Accessed February 15, 2013. http://www.myrepublica.com/portal/index.php?action=news_details&news_id=48900.

Silverman, Phyllis. "An Introduction to Self-Help Groups." In *The Self-Help Group Sourcebook: Your Guide to Community and Online Support Groups,* edited by Barbara J. White and Edward J. Madara , 25-28. Cedar Knolls, NJ: American Self-Help Group Clearinghouse, 2002.

Skinner, B. F. *Science and Human Behavior*. New York: The Free Press, 1953.

Squire, Larry, Darwin Berg, Floyd Bloom, Sascha du Lac, Anirvan Ghosh, and Nicholas Spitzer, eds. *Fundamental Neuroscience*. 3rd ed. Amsterdam: Elsevier, 2008.

Srull, Thomas K. and Robert S. Wyer, Jr. "The Role of Category Accessibility in the Interpretation of Information About Persons: Some Determinants and Implications." *Journal of Personality and Social Psychology* 37, no. 10 (1979): 1160-1672.

Staats, Arthur W. "Unifying Psychology Requires New Infrastructure, Theory, Method, and a Research Agenda." *Review of General Psychology* 3, no. 1 (March 1999): 3-13.

Stone-Romero, Eugene F., Stone, Dianna L., and Salas, Eduardo. "The Influence of Culture on Role Conceptions and Role Behavior in Organisations." *Applied Psychology: An International Review* 52, no. 3 (2003): 328-362.

Sutherland, Max. *Advertising and the Mind of the Consumer: What Works, What Doesn't, and Why*. 3rd ed. rev. Crows Nest, Australia: Allen & Unwin, 2008.

Swing, Edward L., and Craig A. Anderson. "Media Violence and the Development of Aggressive Behavior." In *Criminological Theory: A Life-Course Approach,* edited by Matt DeLisi and Keven M. Beaver, 87-108. Sudbury, MA: Jones and Bartlett, 2011.

Tauber, Robert T. "Good or Bad, What Teachers Expect from Students They Generally Get!" Washington, DC: ERIC Clearinghouse on Teaching and Teacher Education, 1998. ERIC Document Number ED 426 985.

Tedeschi, James T., and Richard B. Felson. *Violence, Aggression, & Coercive Actions.* Washington, DC: American Psychological Association, 1994.

Tharp, Roland G. *Delta Theory and Psychosocial Systems: The Practice of Influence and Change.* Cambridge: Cambridge University Press, 2012.

Thorndike, Edward L. *Animal Intelligence,* 1911. http://psychclassics.yorku.ca/index.htm.

Tipper, Steven P., and Bruce Weaver. "Negative Priming." *Scholarpedia* 3, no. 2 (2008): 4317. Accessed October 1, 2009. http://www.scholarpedia.org/article/Negative_priming.

Toomela, Aaro. "Variables in Psychology: A Critique of Quantitative Psychology." *Integrative Psychological and Behavioral Science* 42, no. 3 (September 2008): 245-265.

Townsend, David J., and Thomas G. Bever. *Sentence Comprehension: The Integration of Habits and Rules.* Cambridge, MA: MIT Press, 2001.

Traynor, Thomas L. "The Impact of State Level Behavioral Regulation on Traffic Fatality Rates." *Journal of Safety Research* 40 (2009): 421-426.

Triandis, Harry C. "Values, Attitudes, and Interpersonal Behavior." In *Nebraska Symposium on Motivation, 1979,* edited by Monte M. Page, 195-259. Lincoln: University of Nebraska Press, 1980.

Triandis, Harry C. *Culture and Social Behavior.* New York: McGraw-Hill, 1994.

Turner, Stephen. *The Social Theory of Practices: Tradition, Tacit Knowledge, and Presuppositions.* Chicago: University of Chicago Press, 1994.

Turner, Stephen. *Brains/Practices/Relativism: Social Theory after Cognitive Science.* Chicago: University of Chicago Press, 2002.

Twain, Mark, quotation. Accessed June 3, 2013. http://www.quoteyquotes.com/quotes_by_author/t/quotes_by_mark_twain_001.html.

Vancouver, Jeffrey B. "The Depth of History and Explanation as Benefit and Bane for Psychological Control Theories." *Journal of Applied Psychology* 90, no. 1 (2005): 38-52.

Vancouver, Jeffrey B. "Self-Regulation in Organizational Settings: A Tale of Two Paradigms." In *Handbook of Self-Regulation*, edited by Monique Boekaerts, Paul R. Pintrich, and Moshe Zeidner, 303-341. San Diego, CA: Elsevier Academic Press, 2005.

VandenBos, Gary R., ed. *APA Dictionary of Psychology*. Washington, DC: American Psychological Association, 2007.

VandenBos, Gary R., ed. *APA Dictionary of Psychology*. 2nd ed. Washington, DC: American Psychological Association, 2015.

Vanderijt, Hetty, and Frans Plooij. *The Wonder Weeks: Eight Predictable, Age-Linked Leaps in Your Baby's Mental Development*. Arnhem, Netherlands: Kiddy World Promotions, 2008.

Vanderstraeten, Raf. "Observing Systems: A Cybernetic Perspective on System/ Environment Relations." *Journal for the Theory of Social Behaviour* 31, no. 3 (September 2001): 297-311.

Vasterling, Jennifer, Richard A. Bryant, and Terence M. Keane, eds. *PTSD and Mild Traumatic Brain Injury*. New York: Guilford Press, 2012.

Verplanken, Bas, Henk Aarts, and Ad van Knippenberg. "Habit, Information Acquisition, and the Process of Making Travel Mode Choices." *European Journal of Social Psychology* 27 (1997): 539-60.

Verplanken Bas, and Henk Aarts. "Habit, Attitude, and Planned Behaviour: Is Habit an Empty Construct or an Interesting Case of Goal-directed Automaticity?." *European Review of Social Psychology* 10, no. 1 (1999): 101-134.

Verplanken, Bas, and Wendy Wood. "Interventions to Break and Create Consumer Habits." *Journal of Public Policy and Marketing* 25, no. 1 (Spring 2006): 90-103.

Vohs, Kathleen D., and Roy F. Baumeister, eds. *Handbook of Self-Regulation: Research, Theory, and Applications*. 2nd ed. New York, NY: Guilford Press, 2011.

von Bertalanffy, Ludwig. *General Systems Theory: Foundations, Development, Applications*. rev. ed. New York: George Braziller, 1968.

von Foerster, Heinz. *Observing Systems*. Seaside, CA: Intersystems Publications, 1984.

von Foerster, Heinz. *Der Anfang von Himmel und Erde hat keinen Namen. Eine Selbsterschaffung in 7 Tagen*, Wein: Döcker Verlag, 1999.

von Foerster, Heinz. *Understanding Understanding: Essays on Cybernetics and Cognition*. New York: Springer-Verlag, 2002.

Wang, Sam. *The Neuroscience of Everyday Life: Course Guidebook*. Chantilly, VA: The Great Courses, 2010.

Watkins, Patti Lou, and George A. Clum, eds. *Handbook of Self-Help Therapies*. New York: Routledge, 2008.

Watson, David L., and Roland G. Tharp. *Self-Directed Behavior: Self-Modification for Personal Adjustment*. 9th ed. Belmont, CA: Wadsworth, 2007.

Watson, David L., and Roland G. Tharp. *Self-Directed Behavior: Self-Modification for Personal Adjustment*. 10th ed. Belmont, CA: Wadsworth, 2014.

Weaver, David R., and Steven M. Reppert. "Circadian Timekeeping." In *Fundamental Neuroscience*. 3rd ed., edited by Larry Squire, Darwin Berg, Floyd Bloom, Sascha du Lac, Anirvan Ghosh, and Nicholas Spitzer, 931-957. Amsterdam: Elsevier, 2008.

Webb, Thomas L., Falko F. Sniehotta, and Susan Michie. "Using Theories of Behaviour Change to Inform Interventions for Addictive Behaviours." *Addiction* 105, no. 11 (November 2010): 1879-1892.

Webster's Ninth New Collegiate Dictionary. Springfield: Merriam-Webster, 1991.

Webster's II: New Riverside University Dictionary. Boston: Riverside Publishing, Houghton Mifflin, 1994.

Weinstein, Neil D. "Testing Four Competing Theories of Health-Protective Behavior." *Health Psychology* 12, no. 4 (July 1993): 324-333.

Wener, Richard. "Effectiveness of the Direct Supervision System of Correctional Design and Management: A Review of the Literature." *Criminal Justice and Behavior* 33, no. 3 (June 2006): 392-410.

Wener, R., W. Frazier, and J. Farbstein. "Three Generations of Evaluation and Design of Correctional Facilities." *Environment and Behavior* 17, no. 1 (January 1985): 71-95.

White, Barbara J., and Edward J. Madara, eds. *The Self-Help Group Sourcebook: Your Guide to Community and Online Support Groups*. 7th ed. Cedar Knolls, NJ: American Self-Help Group Clearinghouse, 2002.

Wicker, Allan W. "Behavior Settings Reconsidered: Temporal Stages, Resources, Internal Dynamics, Context." In *Handbook of Environmental Psychology*, edited by Daniel Stokels and Erwin Altman, 613-654. New York: John Wiley, 1987.

Wilkinson, Richard, and Kate Pickett. *The Spirit Level: Why Greater Equality Makes Societies Stronger*. New York: Bloombury, 2009.

Williams, Kipling D. *Ostracism: The Power of Silence*. New York: Guilford, 2001.

Wilson, Timothy D., and Elizabeth W. Dunn. "Self-Knowledge: Its Limits, Value, and Potential for Improvement." *Annual Review of Psychology* 55 (2004): 493-518.

Wood, Wendy, Jeffrey M. Quinn, and Deborah A. Kashy. "Habits in Everyday Life: Thought, Emotion, and Action." *Journal of Personality and Social Psychology* 83, no. 6 (2002): 1281-1297.

Wood, Wendy, Leona Tam, and Melissa Guerrero Witt. "Changing Circumstances, Disrupting Habits." *Journal of Personality and Social Psychology* 88, no. 6 (2005): 918-933.

World Health Organization. *Assessment of Iodine Deficiency Disorders and Monitoring Their Elimination: A Guide for Programme Managers.* 3rd ed. Geneva: WHO Press, 2007.

Yalom, Irvin D., and Molyn Leszcz. *The Theory and Practice of Group Psychotherapy.* 5th ed. New York: Basic Books, 2005.

Yi, Youjae. "Contextual Priming Effects in Print Advertisements: The Moderating Role of Prior Knowledge." *Journal of Advertising* 22, no. 2 (March 1993): 1-10.

Yin, Henry H. "Restoring Purpose in Behavior." In *Computational and Robotic Models of the Hierarchical Organization of Behavior*, edited by Gianluca Baldassarre and Marco Mirolli, 319-347. Heidelberg: Springer, 2013.

Yue, Xiadong Yue, and Sik Hung Ng. "Filial Obligations and Expectations in China: Current Views from Young and Old People in Beijing." *Asian Journal of Social Psychology* 2, no. 2 (1999): 215-226.

Zajonc, Robert B. "Feeling and Thinking: Closing the Debate Over the Independence of Affect." In *Feeling and Thinking: The Role of Affect in Social Cognition*, edited by Joseph P. Forgas, 31-58. Cambridge: Cambridge University Press, 2000.

Zeiss, Robert, and Antonette Zeiss. *Prolong Your Pleasure.* New York: Pocket Books, 1978.

Zimbardo, Philip. *The Lucifer Effect: Understanding How Good People Turn Evil.* New York: Random House, 2007.

Zimmerman, Wesley W. *The Perception of a Difference: The Power in Buying, Marketing, Selling, Customer Care.* Scottsdale, AZ: WZA, 2005.

INDEX

C

F

G

H

M

N

O

P

R

S

T

U

V

W

Y

Z